ANDREI SAKHAROV
QUARKS AND THE STRUCTURE OF MATTER

ANDREI SAKHAROV
QUARKS AND THE STRUCTURE OF MATTER

Harry J Lipkin

Weizmann Institute of Science, Israel

World Scientific

NEW JERSEY · LONDON · SINGAPORE · BEIJING · SHANGHAI · HONG KONG · TAIPEI · CHENNAI

Published by

World Scientific Publishing Co. Pte. Ltd.

5 Toh Tuck Link, Singapore 596224

USA office: 27 Warren Street, Suite 401-402, Hackensack, NJ 07601

UK office: 57 Shelton Street, Covent Garden, London WC2H 9HE

British Library Cataloguing-in-Publication Data
A catalogue record for this book is available from the British Library.

ANDREI SAKHAROV
Quarks and the Structure of Matter

ISBN 978-981-4407-41-0 (pbk)

Typeset by Stallion Press
Email: enquiries@stallionpress.com

Printed in Singapore by World Scientific Printers.

Contents

Preface   vii

Chapter 1. Quarks and Smuggled Postcards
from Andrei Sakharov   1

Chapter 2. Andrei Sakharov and the Weizmann Institute   19

Chapter 3. The Weizmann Institute and the Scientific
History of Sakharov's Work   29

Chapter 4. How Scientists Study Nature — Pure
and Applied Research   47

Chapter 5. The Building Blocks of Matter — What
is a Quark?   85

Chapter 6. The Forces of Nature   121

Chapter 7. The Weak Force and the Discovery
of the W Particle   137

Index   147

Preface

Andrei Sakharov — a distinguished Russian physicist

Sakharov was the "father of the Russian hydrogen bomb' and also made great contributions to our understanding and use of physics.

He pioneered the search for a way to tame the awesome fusion energy that exploded the hydrogen bomb. He developed methods that could enable fusion energy to be used for peaceful purposes. He also investigated the properties of matter. We knew then that there are two kinds of matter called matter and antimatter. Both kinds were created in laboratory experiments but the universe we knew contained only matter and no antimatter. This included all the stars and galaxies seen by astronomers. Their optical instruments could distinguish between light coming from matter and light from antimatter. They saw only light from matter. The mystery of the absence of antimatter occupied physicists for many years. Sakharov was the first to solve this mystery. Both matter and antimatter were created equally in the "big bang" theory of the origin of the universe. Sakharov showed how all the antimatter could decay leaving only matter. This matter–antimatter problem is still not solved but all searches start with the seminal work of Andrei Sakharov.

Saharov also investigated the structure of matter which was known to made of atoms containing protons, electrons and nuclei. In 1964 the Physicists Gell-Mann and Zweig proposed that the neutrons and protons were really made of smaller constituents called "quarks". The physics community rejected the quark picture as nonsense. But Sakharov and his colleague Zeldovich investigated this quark model and found very interesting results. At the same time this quark model was independently investigated by my group at the Weizmann Institute in Rehovot, Israel. But we did not know of their work until much later.

Sakharov was treated by his government as a great hero until he began to criticize the Russian Government. He wrote critical articles which were circulated abroad. The government then arrested Sakharov and sentenced him to isolation in the city of Gorky which was closed to foreigners. Despite all these restrictions he continued to do scientific work. He was not allowed to publish his work or send it to physicists abroad. In 1980 his wife and friends managed to smuggle some of his work abroad and tried to get it published in the United States. The Americans hesitated because they were afraid that the Russians would claim that Sakharov's work was being published only for political reasons. They wanted to be sure that it was a good scientific paper that would be accepted on purely scientific grounds.

My colleague Tom Ferbel was asked to review Sakharov's paper. Since it was on the quark model of matter and I was a recognized expert at the time, Tom asked me to review the paper. I was amazed to find that it was identical to work I was doing in Rehovot. This led to publicity with international implications. The Russians had claimed that Sakharov was a great man, but he was now old and senile and isolated for his own good. My analysis of his work showed that the Russian excuse was simply wrong. Sakharov was not senile. On the contrary he was engaged in great physics research. The *Washington Post* learned of my analysis and published an article about Sakharov and my work with a title "A voice out of the darkness". This led to a flood of articles in the media and led me to give a series of popular scientific lectures and magazine and newspaper articles for a wide variety of general audiences, including Rotary Clubs, high school students, laymen interested in science, and professors in the social sciences and humanities. These lectures were first motivated by the attempts to help Andrei Sakharov.

My further contacts with the scientific work of Sakharov turned into a remarkable adventure in which I received more manuscripts and postcards from him in his exile in Gorky, smuggled out of the Soviet Union. The story of this adventure and of Sakharov's work and achievements fit beautifully into the description of what scientific research is all about and how it is done. The quarks which were then rejected as nonsense are now recognized as the basic tiny building blocks which are bound together to make the matter we know. Lectures originally given for publicity of Sakharov's plight grew into lectures and articles which fascinated many general audiences, professors and professionals in other fields, and newspaper and magazine editors and readers. They all responded enthusiastically to this combination of the human interest story and achievements of Sakharov with an understandable description of modern physics and the present state of our knowledge of the nature of matter and energy. I explained how the quark building blocks of matter were discovered and understood. I also presented a picture of how how scientists work and develop new knowledge, and discussed the purpose of it all.

Professor Moshe Barasch of the Hebrew University and Mosad Bialik suggested that I put the material together in a book. This book grew out of these lectures.

From earth, air, fire and water to quarks

Throughout the centuries, scientists have puzzled over the questions, what is matter made of, and what are the forces of nature. The ancients believed that everything was composed of four elements, earth, air, fire and water. Now we know that water is a compound made of hydrogen and oxygen and that fire is a chemical reaction that produces energy which can be both useful to mankind and cause terrible destruction. Earth and air are mixtures of many different chemical elements and compounds.

During the twentieth century scientists have made very great progress in understanding the nature of matter and energy. They have found new sources of energy, which like fire are also capable of terrible destruction and of being harnessed to do useful work for mankind.

They have broken up chemical compounds into chemical elements and then found that these elements could be broken up again and again into even smaller objects. They are still investigating whether the tiniest objects known may be made of even tinier building blocks.

In ancient times mankind learned to fear the destructive energy of nature's floods and also to harness the energy of water power for useful purposes. They learned to fear the destructive energy of the wind, and later on to harness it in windmills. Man has learned to use chemical energy in many forms for useful purposes and also to construct explosives for mass destruction. Electric forces and electrical energy were discovered more recently and have become an important part of our everyday life. Nuclear energy in various forms has now been used in recent years both for destruction and to benefit mankind, and there is extensive research today searching for better and safer ways to harness this new kind of fire.

Scientists looking for the basic building blocks of matter found molecules which were made of atoms which in turn were made of nuclei which in turn were made of particles called neutrons and protons. By 1966 new experimental evidence indicated that neutrons and protons were themselves made of even smaller objects. There was a crazy theory proposed in 1964 that the neutron and proton were made of tinier objects called quarks, but nobody had been able to break up a neutron or proton to find these quarks. The majority of physicists believed that the quark theory was plain nonsense.

A few physicists took the quark model seriously and found ways to test this model by experiments on the known particles, even though the quarks themselves were not found and the nature of the forces which held them together in the proton were completely unknown. At the Weizmann Institute in Rehovot, Israel, a group of theorists was active in this field and obtained interesting results which agreed surprisingly well with experiment and provided evidence supporting the quark model. Almost identical work was done independently in Moscow by Andrei Sakharov and his colleague Ya. B. Zeldovich. At that time Sakharov was completely free and his work was published in open scientific journals. But the physics "establishment" did not take quarks seriously at that time and Sakharov's work remained completely unnoticed until fourteen years later.

In 1978, when the quark model had become accepted, I returned to this work in Rehovot and discovered some new relations which agreed remarkably well with experiment. In 1980 I found in Sakharov's smuggled paper that he had independently obtained exactly the same results. This led me to rediscover Sakharov's old results and led to a fascinating correspondence with Sakharov via smuggled postcards.

Who needs quarks?

What is all this quark research good for? What will be its impact on society? Laymen asked the same question about atomic research a hundred years ago. Rutherford then told people that he could not predict the future of atomic research. But he was sure that the government would find a way to tax it.

Now we know that the new understanding of what goes on inside the atom has brought us new technology which has completely transformed our society and our everyday life. Nuclear power plants, transistors, modern computers and lasers did not exist forty years ago. They were developed only after the necessary basic knowledge was provided over a period of years by research into the properties of materials and the interpretation of experiments with the new quantum theory of the atom. Application of that theory to semiconductors enabled the invention of transistors, which in turn made possible the development of modern solid state electronics, which is the basis for modern computers.

This book attempts to describe the search for the basic building blocks of matter and the investigation of the nature of the forces that can either bind them together or blow them apart and can produce either useful or harmful energy. It also attempts to provide a picture of how scientists work and how new scientific knowledge is learned, communicated and passed from one group of investigators to another. The fascinating story of Andrei Sakharov and his contributions to this search provides a simple and easily understood framework for introducing the more technical aspects of the subject.

The first part of the book describes in general terms how scientists study nature and tells the Sakharov story. The second part goes into more detail in explaining the basic building blocks of matter and the development of our knowledge of forces and energy.

Voyages of Discovery

The twentieth century began with the consensus that matter is not continuous but is made up of atoms and molecules. It ended with the confirmation that matter is made of even tinier objects called quarks. But how are such realities established? And how can the process be explained in a plausible way to a non-scientist? I was surprised to find the answer in a book, *The Schools We Need and Why We Don't Have Them* (Doubleday, New York, 1996), by E. D. Hirsch Jr, a professor of English at the University of Virginia. Hirsch's answer also shows the comparison between physical scientific research and pedagogy.

Just as we know that matter is composed of basic building blocks called quarks we also know that literature is composed of basic building blocks called letters of the alphabet. This book describes the voyage of discovery of quarks, the forces that bind quarks into neutrons and protons, and further into the matter that we know. It also describes how to lead a child learning to read on a voyage of discovery of the alphabetic principle, the sounds that bind the alphabet into words and further into the literature that we know. These voyages encounter many obstacles. But the methods of scientific research studying nature have led the way to discovering these obstacles and overcoming them.

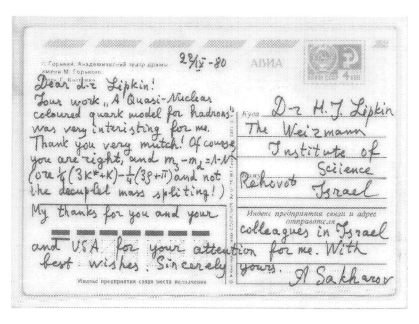

Figure 1.

Дорогой доктор Липкин!

Моя жена Елена и я поздравляем Вас с Новым Годом, желаем счастья Вам и Вашей семье!

Моя жена привезла из Москвы известия, что на радио сообщили, что Вы заинтересовались некой формулой из моей открытки, и я был обрадован этим. Видели ли Вы препринты, изданный в Стенфорде, SLAC-TRANS ~~0191~~ 0191, там есть и другие аналогичные формулы. Среди этих формул - для отношения масс кварков

$$\frac{m_0}{m_s} = \frac{k^* - k}{\xi - \pi} = 1 - \frac{3}{2}\frac{\Sigma - \Lambda}{\Delta - N}$$

и аналогичные формулы для шарма. Массы в этих формулах, конечно, эффективные, с включением части скалярного потенциала конфайнмента. Есть там формулы и другого типа:

$$\frac{\Delta - N}{\xi - \pi} = \frac{1}{2}, \quad \frac{\Sigma - \Sigma}{k^* - k} = \frac{1}{2}$$

и т.п., отражающие ранг цветной группы.

Другие мои препринты этого года - SLAC-TRANS 0190 /"Оценка постоянной взаимодействия кварков с глюонным полем"/ и SLAC-TRANS 0192 /"Космологические модели Вселенной с поворотом стрелы Времени"/.

Научные темы, которые больше всего сейчас волнуют меня - Единая теория поля и Космология ранней Вселенной. Зная, что Вы работаете также в этих областях, я был бы рад иногда обмениваться с Вами письмами /хотя это и очень сложно, ведь формально мне запрещено общение с иностранцами/.

С уважением

A Sakharov

Горький. 9 декабря 1980 г.

Happy New Year!
Best wishes!
Elena Andrei

Figure 2.

1

Quarks and Smuggled Postcards from Andrei Sakharov

In November, 1980 Andrei Sakharov smuggled a postcard out of the Soviet Union from his exile in Gorky. The text of the card shown in Fig. 1 was a response to a reprint of a paper which I had mailed to him. Many people were very excited by this postcard.

It appeared in an editorial in the *Washington Post* and was copied in many newspapers, magazines and journals. Most recently it appeared in the book *The Physicists* by C.P. Snow. It showed that Sakharov was still alert and active in frontier physics despite his isolation, and exposed the rumors spread by the Russians that he was senile and needed to be isolated as completely false. The story behind this postcard is very fascinating.

In the spring of 1980 Professor Tom Ferbel from the University of Rochester Physics Department asked my opinion of a recent manuscript by Sakharov which had been smuggled out of the Soviet Union by a friend. Sakharov was having difficulty getting his work translated and published in the Soviet Union and his friend was investigating the possibility of getting it published in the U.S. He had translated it into English, but he was not an expert in the field of particle physics and did not know whether the work was sufficiently new and interesting to warrant a big fuss about publication. He was in Rochester and consulted Professor Ferbel whom he knew was working in particle physics.

I was at that time on sabbatical in the U.S. and was asked to read and referee the paper to be sure that its scientific value was adequate. The Americans wanted to avoid any charges that they had published the paper for purely political reasons. I was amazed to find that this paper was almost an exact duplicate of work that I had been doing during the past few years.

The history of my contacts with Andrei Sakharov and his association with the Weizmann Institute really begins in 1966, long before we actually met, when he and his Colleague Ya. B. Zeldovich published a paper which referred to four published Weizmann Institute papers: two by H. Harari and H. Lipkin, one by H. Lipkin and one by the Weizmann High Energy experimental group. We did not know of his work at that time, and there is an amusing note added in proof in the paper indicating that he did not know of our ongoing work. The note begins "In a discussion at the summer school in Balaton (Hungary), Bob Sokoloff (Berkeley, USA) developed the hypothesis on the additivity of total cross sections at high energies . . ." Sakharov did not know that Bob Socolow had come to Balaton after spending several months as a visitor at the Weizmann Institute. He reported at Balaton on the work of our group in Rehovot.

That Sakharov's group and our group were doing very much the same work did not come to our attention until 1980, when Tom Ferbel showed me the smuggled paper from the Soviet Union.

I immediately began referring to Sakharov's work in all my lectures and seminars and in invited lectures at conferences and summer schools. In discussions about my own work, I pointed out that the same results had been obtained by Sakharov in Gorky. The reaction of the audience was invariably one of amazement. Prof. Antonino Zichichi, the Director of the International School of Subnuclear Physics at Erice, Sicily was very excited by this. He said it was very important to let the world know that Sakharov

2

was mentally alert and active and not senile. This work of Sakharov should be given maximum publicity. The Russians were circulating rumors that Sakharov was of course a great man, but now he was old and senile and was being isolated in Gorky for his own good. My evidence that this was not true would be crucial to counter this disinformation. The information that Sakharov was able to do any scientific work at all under his difficult conditions of isolation and harassment was very important and should be given the widest possible publicity.

Zichichi ashed me to write a popular article about this work, which he then translated into Italian and published in the Rome newspaper *Il Tempo* with a picture of Sakharov and a headline "An article by Andrei Sakharov from his exile in Gorky". This newspaper was read by many important political figures including the President of the Republic and the Pope. They would see the picture and the headline and understand that the rumor was false. The article appeared in a space regularly devoted to science, whose readers were able to appreciate the scientific aspects of Sakharov's work. But laymen and politicians who also read the paper and only saw the headlines got the message that Sakharov was alert and still making important scientific contributions.

Sakharov was confined to the city of Gorky with no access to scientific institutions nor libraries. The secret police followed him everywhere. They entered his apartment whenever he left and confiscated all papers that they could find. He carried all scientific papers that he needed for his work in his briefcase. Then, one day when he visited the dentist and left his briefcase in the waiting room, two men came in and took the briefcase. Yet he managed to work under these incredible conditions.

It then occurred to me to write Sakharov directly about our common scientific interests. I was assured by friends in contact with him that letters could not possibly do him any harm; he would probably never receive them. But any scientific material that did get through was important, because he felt so isolated in Gorky and appreciated all scientific contacts. Immediately after the Sicily summer school I sent Sakharov a letter and some reprints of my work. He never received the letter but did get the reprint, which duplicated much of his work. He answered with a postcard which he sent somehow to his stepdaughter Tatiana Yankelevich in Boston. She did not know that I was in America, because Sakharov never received my letter. Since the address on the reprint was Weizmann Institute, she naturally sent it to her husband Efrem's brother Boris, who was at that time a doctoral student at the Weizmann Institute. She also enclosed a copy of the smuggled manuscript, not realizing that I had it already. But Boris was doing military service at the time and passed it on to Edward Trifonov, an Institute scientist who knew me, but had no idea what it was all about. He sent the material to me in Chicago with a covering letter about manuscripts from Sakharov. I thought at first that they had gone to a lot of trouble just to send me these manuscripts that I already had. Just as I was about to toss it all out the postcard fell out of the package and I knew that I had received something very important.

At the advice of Sakharov activist Kurt Gottfried, I sent a copy of the postcard with a note to Jessica Matthews of the *Washington Post*. She was then about to write an editorial about the arrest by the KGB of the Soviet "refusenik" scientist Viktor Brailovsky.

At this time many Soviet Jewish citizens had applied for permission to leave the Soviet Union and settle in Israel. When their applications had been refused they sere severely penalized by the Soviet authorities. They were called "refuseniks", fired from their jobs and suffered all kinds of harassment. Refusenik scientists were not allowed to visit scientific institutions. A group of refusenik scientists created their own scientific seminars to enable them to communicate with colleagues and participate in scientific research. At the time that Jessica Matthews received my letter, she saw that it fit beautifully into her editorial.

A few days later my postcard from Sakharov appeared in the leading editorial in the *Post* with the headline "A Voice out of the Darkness". Together with the news of the arrest of Brailovsky, this postcard showed that the KGB could not silence Sakharov and the refuseniks. The postcard was then copied and appeared in many newspapers and magazines, including the *International Herald Tribune*, the *San Francisco Chronicle, New Scientist, Science News*, and C. P. Snow's book *The Physicists*. The story was broadcast by Voice of America in their Russian language broadcast and heard by friends of Sakharov in Moscow who told Elena Bonner who told Sakharov. He then sent me a second postcard, which again made the rounds of the media.

Publicity to Help Sakharov and Refuseniks

Keeping these issues alive and in the media was not easy. After a while the Soviet mistreatment of dissidents and refuseniks was no longer news, and the media were not interested.

Many people in many countries were then working hard to spread the latest news about individual refuseniks, about the harassment and imprisonment of Sakharov, Orlov and Sharansky (SOS) and about the breakup of the Moscow seminars. They also write letters to refuseniks to bolster their morale, to keep them informed of news from Israel and the rest of the world, and to show both the refuseniks and the police who censor their mail that the world had not forgotten their plight and was still actively interested in helping them.

After the arrest of Brailovsky and the breakup of the seminars, groups of concerned scientists in the U.S. organized special seminars similar to those in Moscow with the aim of showing solidarity with the refuseniks and giving publicity to their new troubles. These seminars were held in private homes in university communities throughout the U.S. with a format similar to the Moscow seminar. Outstanding scientists were asked to give talks about their work, and scientists who had attended Moscow seminars as foreign visitors also spoke. In addition there were reports on the present situation and discussions of possible action. Representatives of the press were informed and invited and later wrote articles in the local newspaper. I attended two such seminars, one at the University of Illinois in Urbana and one at the University of Chicago. In both I showed my postcards from Sakharov and told both the scientific and human story behind them.

After my first postcard appeared in the *Washington Post*, I sent the material to the Chicago newspapers, thinking that it might have some local interest. I never received any answer. However, when I sent it to Irv Kupcinet who writes for the *Chicago Sun-Times* and whom I had met at Chicago Weizmann Institute dinners, he was very interested and devoted considerable space to the story in his column. Immediately the competing paper, the *Chicago Tribune*, became interested and sent a photographer out to the laboratory where I was working to take my picture and get a story. But nothing came of this. Local small talk was more interesting for Chicagoans than the story of Sakharov. I also sent a copy of the card and the correspondence to the science editors of the *San Francisco Chronicle* and the *New York Times*. The *San Francisco Chronicle* published a full page in their weekly magazine, with the headline "Courage of an Exiled Scientist" and pictures of Sakharov and the postcard. The *New York Times* did nothing with the story.

In January, 1981 the American Physical Society held a special session on the scientific work of Andrei Sakharov at their annual New York meeting. Several talks summarized Sakharov's contributions to plasma physics and fusion, particle physics, the cosmology of the early universe and the theory of gravitation. A number of these contributions were far ahead of their time and were originally ignored or scoffed at by the physics community. But some of these ideas were accepted over a decade later, when most physicists were unaware of Sakharov's original contribution. Many physicists who had previously known of Sakharov only as the father of the Soviet H bomb and later for his human rights activities were amazed to learn that he had also made very significant contributions to basic research in physics in a wide variety of areas.

After the New York Physical Society meeting, a report on the Sakharov session appeared in the Chicago newspapers, with a picture of Sakharov's step-daughter and a Chicago physicist who spoke there. But there was nothing in the *New York Times*. The story had come from an Associated Press report by their science writer who had been at the meeting. It had been picked up by a number of papers around the country, but not by the *New York Times*. The Associated Press did not report my talk at the press conference about the postcard, because the postcard story was two months old and no longer news. I called the Associated Press reporter after receiving the second postcard. He immediately wrote a story which appeared in many newspapers, with headlines like "Sakharov Finds Way to Beam His Idea" and "Notes from the underground: Soviet smuggles cards". Voice of America was very interested in the story that Sakharov had heard of my interest in his work from VOA. They offered to broadcast a response to his second postcard which I dictated and they translated into Russian. But major newspapers like the *New York Times* and the *Washington Post* did not carry the story. Sakharov's difficulties were old hat and not considered news any more.

I attempted to get as much publicity as possible for Sakharov in the media by the use of the two postcards. By this time a story had unfolded that could be told in many different ways to many different audiences from scientific seminars to rotary club meetings. Different versions of the story appeared in a number of publications.

I followed up the appearance of my first card in the British magazine *New Scientist* by writing to the Editors expressing strong approval about their use of my postcard. But I was surprised that they had not informed me. Their response was an apology and a request that I write an article about Sakharov's scientific work. By this time I had been impressed with the breadth and depth of his work and responded that he had done so much in so many different fields that no one person could adequately cover it all.

I agreed to write an article about my field and suggested that there be two other articles about his other contributions. The result was a special section of *New Scientist* on Sakharov's work, timed to correspond with a meeting in New York celebrating Sakharov's sixtieth birthday. It contained the three articles following the general outline of the New York Physical Society meeting, an editorial on Sakharov, entitled "An honorable dissident", with a picture of Sakharov on the cover of the magazine.

This section appeared in the April 30 issue, just in time for the international meeting in May 1981 honoring Sakharov's 60th birthday in New York. The program featured his contributions to Science, War and Peace and Human Rights. The reports by outstanding speakers in all these areas were very impressive. In all three areas Sakharov has been ahead of his time.

He had not been a lone naive voice in the wilderness fighting for hopeless causes. His contributions can be characterized as combining brilliant vision and analysis with a down-to-earth practical view of what will be realistically feasible ten years hence, and as recognizing the significance of new developments long before they were fully appreciated by others. He had already lived to see some of his early ideas which were ridiculed at the time accepted ten or fifteen years later.

The "Moscow Seminars in Exile" in Urbana and Chicago attracted the attention of the press. The Champaign-Urbana paper carried a big story, including material on Sakharov and the postcards. The two Chicago papers carried stories which mainly emphasized the presence of Sakharov's stepdaughter Tatyana Yankelevich who had come specially from Boston for the seminar. The plight of the refuseniks was also mentioned, and of course the fact that one of the speakers was a Nobel Prize-winner helped get attention from the press.

The story of my correspondence with Sakharov finally made the *New York Times* after the international symposium honoring his 60th birthday, as part of a report on the meeting. The meeting also provided an occasion for the story to appear in newspapers which had been sitting on it for some time, waiting for a suitable occasion.

The story of the postcards meandered through the media in unexpected ways. Direct efforts to send the story to editors usually were unsuccessful, unless the particular writer had a special interest or a special incentive to write about it. But the public affairs office of the Argonne Laboratory, which had done nothing to publicize the story, noted the article on my talk at the Urbana Moscow Seminar in the local paper and assigned a summer student working at Argonne to look into it. She interviewed me and wrote a very good article entitled "Sakharov Writes to Argonne Scientist — East meets West through smuggled postcard", which was published in the internal Argonne magazine, *Argonne News*. This story was then picked up by another student, working for a master's degree in journalism at a local university, who called and asked for an interview. She thought the story would be suitable for the local Chicago papers. I was skeptical, since the direct contact with the editors and other reporters had led nowhere. But shortly afterwards she called, said that she was now working for the *Chicago Sun-Times*, and was ready to use the story. She sent a photographer to take shots of me with the postcard, and a full page article appeared shorly afterwards. Never underestimate the power of a student.

But by the fall of 1981, Sakharov was almost a forgotten man in Gorky. The Russians were succeeding in isolating him and keeping his views from reaching the Soviet people and the world. His letters received little attention when they were smuggled out of the Soviet Union. He sent a long letter on the suppression of human rights and on his own condition to a well-known physics professor in America. But no major newspaper nor magazine would publish it, because the standard suppression of human rights in the Soviet Union and the mistreatment of Sakharov was no longer news. The Soviet strategy was clear. Wait until the West gets tired of all this dissident and refusenik business. Then the KGB will be able to do as they please without notice or interference from abroad.

Sakharov found the way to get back into the media and the headlines. His wife's two children by her former marriage had emigrated to the United States. They had originally not wanted to emigrate, but they had been harrassed so much by the KGB because their mother had married Sakharov that they finally applied for emigration. But Sakharov's stepson's fiancee, Lisa, was still in the Soviet Union and was not allowed to leave and join her fiance. Sakharov and his wife went on a hunger strike to force the authorities to let her go.

When Sakharov began his hunger strike, many of his friends in America were very critical. Why make all this fuss about one girl who couldn't get out when so many refuseniks like Anatoly Sharansky are in much worse condition? Why is Sakharov going overboard for his own family? They did not appreciate Sakharov's brilliant tactics in bringing the whole problem of Soviet dissidents and refuseniks very dramatically to the attention of the world.

Strong letters of protest against the treatment of Anatoly Sharansky or even a hunger strike for Sharansky would have received very little publicity at that time. The world and the media were bored with the problem of Soviet dissidents. But the media went wild about the human interest story of this poor girl who only wanted to join her fiance, and whose only crime had been to want to marry a man whose only crime had been to let his mother marry Sakharov after he was already a grown man. The Soviets were trapped. They were besieged from all directions with the simple question "Why don't you let this poor girl go?" and they had no answer. They couldn't even find a standard Soviet lie, like calling Sharansky a CIA agent. Whenever their representatives abroad wanted to talk about important things to important people, they found themselves having to give evasive answers to the question "Why don't you let this poor girl go?"

Many people dismissed the possibility of exerting pressure on the Soviets by saying that it only depended upon getting Reagan to put pressure on top Soviet leaders like Brezhnev, Andropov or Chernenko. But pressure could have been exerted at all levels in the Soviet hierarchy. The Russians had other problems besides

Sakharov and Soviet Jewry, and many were much more important and had higher priority. We exerted pressure whenever we could show the Russians that they were losing something more important by harassing Sakharov or Soviet Jews.

Sakharov's hunger strike was a beautiful example of such pressure. Brezhnev was going to West Germany to meet with Chancellor Schmidt. He wanted the maximum publicity for his visit, for his propaganda messages of peace and security and for his arguments against placing American missiles in Europe. The last thing he needed was to have his visit obscured by headlines in the media about Sakharov's hunger strike while Brezhnev's peace talk would barely make the back pages and to have all press conferences highlighting the question of why he did not allow Sakharov's stepson's wife to go to America. Sakharov won. They let Lisa join her husband. But Sakharov also won much more.

Sakharov's tactics showed a deep understanding of how the Soviet system works and of the psychology and the media in the West. His victory went far beyond the simple liberation of his daughter-in-law. He lost no time in using the limelight in the media to call the world's attention again to the despicable treatment of human rights in the Soviet Union and to give more publicity to the plight of dissidents and refuseniks in trouble.

During Sakharov's hunger strike his relatives here in Israel asked me to help organize a demonstration against Soviet Nobel Laureate Nikolai G. Basov, an outspoken leader in the denunciations of Sakharov, who was about to attend an International Conference on Lasers in the United States in December 1981. I did not find any Israelis going to that conference, but sent telegrams telling of Basov's visit to friends including Prof. Morris Pripstein of the University of California at Berkeley, chairman of SOS (Scientists for Sakharov, Orlov and Sharansky).

Prof. Pripstein did not know anyone attending the laser conference, but found out that Basov was attending another conference in San Francisco. A demonstration by prominent American scientists including one Nobel Laureate was described by large articles with pictures in the two major newspapers and by smaller articles in six other local papers. He sent copies to me with a letter: "Enclosed are the press clippings of our demonstration against Basov which we organized as a result of your telex to us. We didn't know he was coming since the organizers of the conference in San Francisco tried to keep it quiet. We learned from sources who were in contact with Basov afterwards that he was extremely upset about our demonstration."

In 1983 Andrei Sakharov was still sitting in Gorky, held without trial and without being accused of any crime. He continued to send his messages very clearly to the world, and we wondered how we could do anything to help. The least is to try to maintain all possible contact to help him in his isolation, and to let the rest of the world and the Russians know that we have not forgotten this great man. The Soviet authorities undoubtedly wished that everyone would forget about Sakharov, the refuseniks and the dissidents, so that they could get rid of all of them quietly. We did our best to prevent this scenario.

The Harassment of Andrei Sakharov
Despite His Scientific Achievements

In October 1983 I passed through Boston, called the Sakharov family and spoke to his stepson Alexei. He was now happily reunited with his wife who had been allowed to join him as a result of the hunger strike. But he told me that Sakharov's situation had become worse because the KGB had finally succeeded in inciting the general public against him after his article expressing his views on how to end the nuclear arms race was published in the American magazine *Foreign Affairs*. Still fighting for his views, with no regard for its impact on his own situation, he had smuggled this article out of the Soviet Union. The KGB propaganda called him a warmonger and aroused the public. When one recalls how the antiwar people in the U.S. are aroused against Teller it is easy to see how the KGB controlling the media can give Sakharov, the father of the Soviet H-bomb, the image of a superhawk who wants to destroy the world in a nuclear war, even though he is fighting for nuclear disarmament and preventing nuclear war. The Sakharovs were now receiving threatening phone calls and were harassed by people they met on the street.

Meanwhile both Andrei and Elena Sakharov needed hospitalization and medical treatment. The Soviet authorities refused to allow them both to be hospitalized in Moscow as recommended by the doctors of the Soviet Academy of Sciences. Instead they were hounded by the KGB who followed him everywhere, confiscated all papers he left in his apartment, and again stole the papers he needed for his work, this time by attacking him bodily in his car. Elena had been told that she could be treated in the Academy hospital in Moscow, but she refused to go without Andrei. Andrei refused to go to the hospital in Gorky because he did not trust them after his treatment there during the hunger strike. They gave him various medications without telling him what they were.

Despite his illness, harassment and confinement in the city of Gorky, where he was barred from any access to scientific institutions or libraries, he continued to work and to be heard. He carried on trying to do physics and was writing another paper on the matter–antimatter problem in the early universe which was almost ready to be submitted to the prestigious Soviet *Journal of Experimental and Theoretical Physics* (JETP).

When I mentioned that there were rumors of a proton decay event in Japan, Alexei asked me if anyone had written to Sakharov about it. He said that Sakharov was starved for news about physics, and any news of this kind would be very good for his morale. Letters could be sent both to his address in Gorky and to the Lebedev Institute, where his wife picked up the mail. Some got through.

I immediately wrote to Professor Larry Sulak at the University of Michigan who was then active in a proton decay experiment and had been very excited about the proton decay rumor, suggesting that he write to Sakharov. When I returned home I wrote to others who had been active for Sakharov and active in the field of proton decay, urging them to keep Sakharov up to date by writing him letters and sending him copies not only of published papers but also interim reports on the progress of experiments, proposals for new experiments, letters to professional colleagues discussing the latest results and rumors, etc. I then suggested the same for all scientific refuseniks, who are so completely isolated from the everyday excitement of scientific research. I later received an answer from Prof. Sulak, thanking me for the suggestion and saying that he had never thought of sending anything to Sakharov before because he was sure that they would never get through.

Meanwhile the official campaign against Sakharov openly stated that he was mentally ill, and gave rise to fears that there would be attempts to place him in a mental institution. Soviet spokesmen up to the top level of Yuri Andropov told foreign statesmen and the press that Sakharov was unbalanced, even using the word "cuckoo" and needed isolation for his own good. The *Washington Post* again used the stories of my contacts with Sakharov in an article to refute these accusations. Robert Kaiser, who had spent three years in the Soviet Union and knew and respected Sakharov personally, was now an editor of the *Post*. He saw the article I had written for Sakharov's honorary doctorate at the Weizmann Institute. At his request I rewrote it as an answer to the Soviet propaganda statements. It appeared in a Sunday *Washington Post* with the headline "It's not Sakharov who's Cuckoo" and was reprinted in a number of other major newspapers, including the *International Herald Tribune* and the *Manchester Guardian Weekly*.

I do not know how much these demonstrations and newspaper publicity really helped. But one of my prized possessions is a photograph of Andrei Sakharov taken in Gorky, with the inscription "to Harry, true friend, from Sakharov's family."

Continuing Harassment of Andrei Sakharov

Appeals for Humane Treatment of Sakharov

The lone voice of Andrei Sakharov's appeals for human rights and control of nuclear weapons had repeatedly attracted the attention of the world. As a brilliant scientist at the very top of the Soviet elite, he could have had everything he wanted before he gave it all up to fight for freedom and human rights. His courage, wisdom and strength were an inspiration to the Soviet dissident movement and to the Soviet Jews fighting for the right to leave the Soviet Union and come to Israel. Although he was not Jewish himself, he always supported the Soviet Jewish refuseniks in their struggle. Before his exile to Gorky, he attended the scientific seminars organized by the refuseniks.

Andrei Sakharov was an outstanding scientist. His contributions to science, as well as to war and peace and to human rights included early ideas which were ridiculed at the time and accepted ten or fifteen years later. A century from now, when all the political problems of today have turned into ancient history, students may still be studying his explanation of the absence of antimatter in the universe and his contributions to the development of fusion energy. He might even be considered as a Russian Galileo, whose important research contributions survived despite the harassment by the political powers of the time.

History has passed its judgement on the Catholic church for its harassment of Galileo and on the U.S. Government for its harassment of J. Robert Oppenheimer. In my letters published in major newspapers I reminded the Soviet leaders that they could face a similar judgement. Their grandchildren would some day study the work of this great Russian physicist and wonder with embarassment about the role of their grandfathers in his persecution. The Soviet authorities did not succeed in rewriting history to erase Sakharov's achievements. The western scientific textbooks and encyclopedias continue to describe his contributions to the research which enabled mankind to understand and control the forces of nature. Perhaps arguments like these did influence the Soviet authorities to stop the senseless harassment and persecution of one of their own great men.

Despite his isolation and harassment, Sakharov continued his scientific work as well as his struggle for more freedom in the Soviet Union and for ending the nuclear arms race. His efforts were limited to those areas where he could make useful contributions despite his minimal contact with the outside world. Fortunately he anticipated much of today's research with his pioneering work in the 1960's. The scraps of information that he received in Gorky were sufficient for him to see how his 1966 ideas could be developed further in the light of recent developments. The papers that he had written and still wrote in Gorky described such development of his work on the quark theory of matter and on the cosmology of the early universe.

The scientists all over the world engaged in investigating the structure of matter, looking for proton decay and developing new theories for the early universe were denied the benefit of contacts with Sakharov. One can only wonder what the impact on their work would be if Sakharov had been free to visit their laboratories, learn about their latest results and future programs and give them the benefit of his great knowledge and insight. It is a disgrace that a scientist of Sakharov's stature was not allowed to pursue his work for the benefit of all mankind with the freedom which we enjoy in the West.

Sakharov wrote letters and scientific papers which somehow managed to get out of the Soviet Union and were published abroad. The scientific value of these papers has also been recognized by his Soviet peers, who received them and allowed them to be published in their own prestigious Soviet *Journal of Experimental and Theoretical Physics*. Soviet scientists who read their own journals knew that Sakharov was still alert and active in frontier research, and not "unbalanced and isolated for his own good" as the government authorities claimed. His 1983 letter about the nuclear arms race published in the magazine *Foreign Affairs* also showed clear incisive thinking. In Gorky he was still pushing ideas whose time had not yet come. Freedom in the Soviet Union and an end to the nuclear arms race seemed like wild dreams. Now we can see that Andrei Sakharov was right.

2

Andrei Sakharov and the Weizmann Institute

Sakharov's Contributions and Achievements

In May 1983 the Weizmann Institute of Science recognized the many achievements of Sakharov by awarding him an honorary doctorate *in absentia*. I was asked to receive the degree for him. Sakharov was known for his struggle for human rights in the Soviet Union and had already been awarded the Nobel Peace Prize. But he was being recognized here also as an outstanding scientist whose contributions rank among the leading pieces of research in theoretical physics of the past thirty years.

Sakharov's family was represented at the honorary degree ceremony at the Weizmann Institute by his stepdaughter's mother-in-law, Tamar Yankelevich, who lives in Israel. She brought me another postcard from Sakharov, a New Year greeting, together with a volume of selected reprints of Sakharov's scientific papers, autographed by her son on behalf of Sakharov. I also wrote an article emphasizing Sakharov's scientific stature which was distributed on the occasion of the award, and sent copies to major newspapers both in Israel and abroad. At that time the press was not interested.

Sakharov began working on the Soviet atomic bomb project in 1948 shortly after he began his graduate studies in physics. He is generally considered to be the father of the Soviet hydrogen bomb, and has been characterized as the Soviet equivalent of Robert

19

Oppenheimer, Edward Teller and Hans Bethe, all rolled into one. At that time he felt that it was important for both superpowers to have the bomb, because then no one would dare to use it again like in Hiroshima and Nagasaki. But once they had it, he became increasingly concerned with the moral and environmental implications of nuclear testing and industrial pollution.

Sakharov's contributions to the political aspects of science were also ahead of their time. He saw clearly the necessity for a nuclear test ban at a very early stage. He also saw that banning only tests in the atmosphere was the only practical first step. In the late 1950's Sakharov began a campaign to halt the testing of nuclear weapons. Negotiations between the Americans and the Russians on nuclear testing were stalled. The Americans mistrusted the Russians and insisted on the right of inspection inside the Soviet Union. The Russians mistrusted the Americans and feared that the Americans would use inspection as a cover for espionage. A realistic visionary, seeing clearly the crucial points ahead of his time, Sakharov noted that the inspection issue only affected underground testing which would not cause radioactive fallout and pollution. He proposed a first step banning only the dangerous tests in the atmosphere, under the sea and in outer space. These could easily be detected without inspection. He was not taken seriously, came into conflict with the highest authorities including then Soviet Premier Nikita Khruschev, but persisted in his efforts and helped to promote the 1963 treaty banning just these tests. This test ban, successfully observed to this day, is still the main effective international agreement protecting mankind from the dangers of radioactive fallout.

Already while working on the H-bomb, Sakharov began investigating the peaceful uses of this awesome energy source, which is called nuclear fusion, and is most simply described as a new kind of fire which is literally too hot to handle. We can use the energy from burning oil and coal because we have materials like steel to build furnaces and stoves that can contain the fire. But the fire of nuclear fusion is at a very high temperature that no known material can survive. It is as if we had no metals and had only plastics to build stoves and furnaces for burning oil. This is why it is easier to use fusion for bombs than for peaceful energy. To make a bomb you don't need a furnace to hold the fire; you light it once and let it blow up.

In 1950 Andrei Sakharov and Igor Tamm wrote the fundamental scientific paper of the Soviet research effort on controlled fusion, in which they proposed containing the hot fusion fire in a "magnetic bottle". Powerful electromagnetic forces could prevent the hot fire from reaching the walls of the container and allow the fire to keep on burning without destroying the furnace. Sakharov developed a type of magnetic bottle called the "tokamak" which is today the main line of international research on fusion reactors. If Sakharov's "tokamak" can be developed into a new kind of furnace which answers the energy crisis, Sakharov will be remembered with the long list of physicists whose researches delivered new energy sources to mankind.

In 1966 Sakharov wrote two remarkable papers in particle physics very much ahead of his time. Throughout the centuries, scientists have puzzled over the question: what is matter made of? They found molecules which were made of atoms which in turn were made of nuclei which in turn were made of particles called neutrons and protons. By 1966 new experimental evidence indicated that the neutrons and protons were themselves made of even smaller objects. There was a crazy theory proposed in 1964 that the neutron and proton were made of tinier objects called quarks, but nobody had been able to break up a neutron or proton to find these quarks.

Sakharov and his colleague Ya. B. Zeldovich took the quark model seriously and found ways to test this model by experiments on the known particles, even though the quarks themselves were not found and the nature of the forces which held them together in the proton were completely unknown. Almost identical work was done independently at the Weizmann Institute by Pedro Federman, Hector Rubinstein and Igal Talmi. The results they obtained agreed surprisingly well with experiment and provided evidence supporting the quark model. But the physics "establishment" did not take quarks seriously at that time and Sakharov's work remained completely unnoticed until fourteen years later.

In 1978, when the quark model had become accepted, I returned to this work in Rehovot and discovered some new relations which agreed remarkably well with experiment. These were the results which had also been independently obtained by Sakharov and reported in the paper smuggled out of the Soviet Union in 1980.

Even more remarkable was Sakharov's explanation in 1966 of the absence of antimatter in the universe; a serious problem which had baffled the cosmologists. The protons and neutrons have their "antiparticles" called antiprotons and antineutrons. These can combine to make antinuclei, antiatoms and antimatter. Such antimatter can be produced in laboratory experiments and scientists wondered whether there were stars and planets made of antimatter somewhere in the universe. In the accepted "big bang" theory of the origin of the universe, there must have been equal amounts of matter and antimatter created, and there were speculations that there must be stars and planets made of antimatter somewhere in the universe. There were even jokes about Prof. Edward Anti-Teller who worked in an anti-laboratory on an anti-planet where all of the theories of physics were developed by anti-semites. But all the astronomical observations failed to find any trace of antimatter.

All other astrophysical observations favored this theory of the origin of the universe in which everything started with a "big bang" that created equal amounts of matter and antimatter. But there was no known mechanism for getting rid of all this antimatter. The astronomers' searches should have found some trace of the antimatter created in the big bang if it were still around. Thus there seemed to be a contradiction between the big bang theory and the astronomical observations.

Sakharov showed in 1966 how all the antimatter created in the big bang could have decayed away, leaving only matter. This resolved the contradiction with the big bang theory. But Sakharov's theory was not accepted because he made a number of assumptions and predictions that were ridiculed as being completely crazy at the time. He predicted that protons should decay into electrons and other particles and everyone knew that protons did not decay.

Today Sakharov's theory has become the accepted view on the matter–antimatter problem. His crazy assumptions are now a central part of the standard theory. Large expensive experiments all over the world are looking for proton decay predicted by Sakharov. Recently a Japanese group searching for proton decay reported some events in their experiments which look like signals from protons decaying, but their results are not yet conclusive and need further verification. This could be the most exciting discovery of the century. Most physicists are amazed that anyone could have foreseen all this in 1966.

In May, 1990 shortly before my wife Malka and I flew to New York from Israel, Boris and Larissa Altshuler, friends of Sakharov, visited us in Israel and gave us some manuscripts to give to Ed Kline in New York. We did not know Ed Kline, and they did not know his address or telephone number. But they said that Elena Bonner would know, and that she was visiting her son Alexei in Boston at that time. The telephone number that they gave us for Alexei turned out to be wrong (two digits were interchanged) and I had no idea how to find an Ed Kline (Klein? Cline?) in New York.

But the physics mafia works in strange ways. Prof. Sidney Drell, a good friend of the Sakharovs, was spending a sabbatical from Stanford at Rockefeller University in New York, which I was also visiting. I met Sid and asked if he knew someone named Ed Klein. "Of course", he said. "He is a wonderful person, a business man who is interested in human rights and has looked after the Sakharov children ever since their arrival. By the way, Elena Bonner is in New York today, and Ed Kline will know how to contact her."

Sid and I had a long talk about Sakharov. During Sakharov's last visit to the U.S. Sid had asked him whether it wasn't strange that this great breakthrough toward reform came from Gorbachev, a man who had been close to the KGB all the time and would be expected to support repression. "No," said Andrei, "it is not strange at all. The KGB are the only people in the Soviet Union who have all the information about how bad the situation really is. Only they know that drastic reform is necessary to prevent everything from falling apart."

This little remark gives some insight into Sakharov's brilliant abilities and talents. He was not only a great fighter for human rights and a man of high moral stature. He understood the functioning of the system repressing human rights and saw its weak points better than the people who were running the system. He also understood that the system would not be changed only by moral arguments; it was necessary to convince the leaders that the system would not work unless these changes were made.

It is interesting to apply this insight to the cause of Andrei's exile to Gorky: the Soviet invasion of Afghanistan. This was not only morally wrong, it was doomed to failure. If the leaders could have been convinced that it would fail, they would not have invaded Afghanistan. But as long as they believed that it would succeed, no moral argument or international pressure would stop them.

Andrei Dimitriyevich had the remarkable ability to understand how systems function — all systems, social, political, scientific, technological, as well as the interfaces between them. In a sense he was a kind of an interdisciplinary systems super-engineer. I now realize that during the process of working for his cause I learned many things from him. One example of his understanding of systems from which I learned a great deal was exhibited by his undertaking his hunger strike in Gorky.

I recently gave some of my old articles to friends of Sakharov visiting Israel. They were particularly impressed by one article published in the *Jerusalem Post* whose main thrust was to show how it was possible to work for the causes of refuseniks in the USSR and stressed what we could learn from Sakharov and in particular from his hunger strike. At that time many people dismissed the possibility of exerting pressure on the Soviets by saying that it only depended upon getting Reagan to put pressure on Andropov. I noted that the Russians have other problems besides Soviet Jewry, and many are much more important and have higher priority. Pressure is possible by showing the Russians that harassing Soviet Jews makes them lose something more important. I presented the hunger strike of Andrei Sakharov as a beautiful example of effective pressure.

That Sakharov's group and our group were doing very much the same work did not come to our attention until 1980, when Sakharov was in exile in Gorky and Tom Ferbel showed me the manuscript smuggled out of the Soviet Union.

Thus began a long campaign to publicize the scientific work of Sakharov. His human rights activities and his harassment by the KGB were well known and well publicized by others. But the importance and implications of his scientific work were not well known and offered us the opportunity to do something that wasn't already being done better by others. There were three directions to this public relations effort:

1. To show that he was still active in science and that rumors about his senility were false.
2. To tell an interesting story about his scientific work and his contacts with the Weizmann Institute. Such stories could find new places in the media and help keep Sakharov's problems in the public view. After a short time, the public and media were bored with hearing one story after another about human rights and Sakharov's ordeals in Gorky. It was hard to get news about Sakharov into the media. The scientific angle kept offering new opportunities for interesting articles.
3. The Galileo–Oppenheimer comparison. "Today the world remembers the names of Galileo and J. Robert Oppenheimer, while the names of those who persecuted them are forgotten". Sakharov's scientific contributions were sufficiently important to remain in a history that Brezhnev and Andropov would not be able to erase. They should wonder what their grandchildren will think when they study the work of this great scientist in school.

An article written for the Weizmann Institute magazine *Rehovot* attracted the attention of the editor of the *London Times Higher Education Supplement*, who asked me to write an article along the same lines for them. In 1983 I used the Galileo theme in an article requested by the *Washington Post* which was then also printed by other newspapers including the *Manchester Guardian*. In 1984 I used this theme in an article published in the *Guardian* suggesting the creation of a prestigious Andrei Sakharov Prize for Energy Research where I pointed out that Sakharov's contributions in this area justified a prize bearing his name.

In 1983 the Weizmann Institute awarded Sakharov an honorary doctorate, and I was asked to receive the degree for him at the ceremony. Here again we used the occasion to publicize Skaharov's plight and to emphasize that he was a great scientist as well as a hero fighting for human rights. In 1985 I told my Sakharov story to an audience of biologists as part of the traditional interdisciplinary session of a conference in France where a physicist tries to tell about recent events in physics in a language understandable to biologists. The "Sakharov–Zeldovich" model figured prominently in talks given at a number of conferences and summer institutes.

After Chernobyl, I wrote several articles pointing out that if Gorbachev was really serious about working to prevent further nuclear accidents, he should get the best brains in the field of nuclear safety to work on this problem. The first step is to send one of the best scientists in this field, Andrei Sakharov, back from Gorky to Moscow where he can work again. In 1986, shortly before Sakharov's release from exile, I mentioned his work in a talk given at an international conference held in the USSR. I had checked with my hosts who told me that I was free to mention Sakharov's scientific work, as long as I didn't say anything political. Knowing the Soviet cultural trait of reading between lines, I began my talk by recognizing the great Soviet contributions to the field of my talk. I then expressed my regrets at the absence from the conference of some physicists who had made great contributions and whose participation in the conference would have been extremely useful.

Sakharov's return to Moscow ended this chapter in our public relations effort to promote his case. We actually met in Moscow in August, 1988, when we had dinner together at the apartment of Arkady Migdal, and were together at the Pugwash Conference in September, staying at the same hotel in Dagomys. I did not realize that this would be the last time that I would see Andrei, and we were looking forward to a visit by Andrei and Elena in 1990. Instead, I met Elena in Jerusalem in June 1990 at the dedication ceremony of the Andrei Sakharov Gardens in Jerusalem and at a ceremony at the Weizmann Institute where we had hoped to give Andrei his honorary degree in person. But Andrei was no longer with us.

The Jerusalem ceremony was a unique occasion honoring Sakharov. He will be given many honors by people and organizations all over the world who recognize his greatness and struggle for human rights. The Jerusalem ceremony was different. The people there were also giving a very personal message of thanks to Sakharov for his help and support in their own individual struggles for freedom and the right to leave the Soviet Union. They were now free and living new lives and wanted to thank Sakharov for his hard work in their behalf. The Sakharov Gardens are not only a memorial to a great man. They also represent the thanks of a multitude of people who were helped by him. I spoke at the dedication ceremony and was introduced as a "friend of the Sakharov family", who had asked that I be given the opportunity to say a few words. I was very touched by this introduction, and by being told by Elena and Tanya afterwards "Of course you and Malka are our friends, and Andrei always spoke of you in this way."

At the Weizmann Institute ceremony, one of Sakharov's unique contributions to the struggle of the Jewish refuseniks was described by the Minister of Science and Energy, Prof. Yuval Ne'eman, who told how Sakharov had once telephoned him personally to inform him that Evgeni Levich, the son of the refusenik Academician Benjamin Levich, had been called up for army service in the Arctic and that this was done only to harass the family and had no connection with Soviet security. Sakharov asked Ne'eman to publicize this case and his statement in the West as a way to bring pressure on the authorities.

Sakharov's friends told me just recently that his hunger strike is still being criticized by many people who do not understand the point that he was making and which I stressed in my article. They urged me to include my explanation of the real nature of the hunger strike in an article for the Sakharov Memorial Volume. This explanation is also important because there is also a campaign by forces opposing perestroika to paint Sakharov as a great scientist who was led astray by his "Jewish" wife into anti-Soviet activities mainly to help her family. The hunger strike is given as an example of such activities.

I conclude this chapter by returning to the story of Ed Kline and Elena Bonner in New York. I called Ed Kline at the number that Sid Drell had given me. That evening Malka and I had a wonderful visit at the Klines' apartment with Elena Bonner who was spending the night there. We discussed the hunger strike and its motivation and significance again. Elena Bonner told us that the evening after Lisa received her permission and Andrei ended his hunger strike, a group of friends including Sharansky's mother Ida Petrovna were celebrating with her in her apartment in Moscow. Someone criticized her for making such a fuss about this girl in her own family in front of the mother of someone who is in prison. Ida Petrovna retorted sharply Andrei is fighting for all of us!

Andrei is gone. But he still lives on in our minds and our hearts. And his heritage is still fighting for all of us!

3

The Weizmann Institute and the Scientific History of Sakharov's Work

While publicizing Sakharov's recent work in Gorky, I also investigated the scientific history of this work. The text of the first postcard raises several interesting questions. My paper to which he referred was thirty pages long with many formulas. Why did he select this one particular formula for his comment? His remark "Of course you are right and ..." suggest that my equation is controversial, and that he supports my argument. Why did he pick this particular controversial point? And above all, how did he happen to do exactly the same work in Gorky that I was doing in Rehovot? The answers to these questions are very illuminating and give interesting insight into the methodology and communication of scientific research.

Following up the Formula in the Postcard

Following up the implications of Sakharov's postcard had all the excitement of a good detective story. The manuscript smuggled out of the Soviet Union in 1980 referred to a previous paper by Sakharov and his colleague Zeldovich, published in the *Soviet Journal of Nuclear Physics* in 1966. This paper was available in any physics library in the West, but had been ignored and was not known. The paper contained work almost identical to work done at the Weizmann Institute at the same time by my colleagues Hector Rubinstein, Igal Talmi and Pedro Federman. This coincidence is not surprising. These groups in Moscow and Rehovot were both strong in nuclear physics and not part of the particle physics establishment at that time. It was very natural for theorists with this background to use the techniques of nuclear physics to describe the structure of the new particles.

The main part of the Sakharov–Zeldovich paper did not duplicate the work of the Rehovot group, but proposed a new model to explain some recently discovered new particles at the accelerator at Berkeley, California. These particles had an electric charge three times the charge of the electron and could not be accommodated in the accepted models. Unfortunately this discovery, although published in the scientific journals and presented at an international conference in particle physics, turned out to be premature. Continued experimental work showed that there were no such particles with a charge three times the charge of the electron.

This excitement about premature results occurs frequently in frontier research, where experiments are difficult, and preliminary indications of new and unexpected exciting phenomena are often unreliable. The experimenters cannot verify the existence of these new effects immediately; it requires the design of a new experiment, approval of the budget by a funding agency, and the granting of precious time on an accelerator by the program committee. But meanwhile the news leaks out to the theorists, who immediately start thinking about explanations for these results, just in case they should turn out to be right. If the new effects disappear in the next generation of experiments, they are quickly forgotten, and the theoretical papers explaining them sink into oblivion. This was the fate of the 1966 paper by Sakharov and Zeldovich.

This forgotten paper contained in a small corner the same results obtained by the Rehovot group. They were valid in the standard model and independent of the additional assumptions introduced to explain the nonexistent new particles. It also contained one result which the Rehovot group had not obtained, namely the formula which appears on Sakharov's first postcard in Fig. 1. I discovered this result much later, in 1978, and was very excited by it because it was a simple prediction from the model which agreed surprisingly well with the available experimental data. I was sure that someone else must have found it earlier, but checked both by looking in the literature and asking all the experts. Nobody had seen this result before, so I published it as original.

Sakharov was very generous in his postcard, writing "Of course you are right, and . . ." before this formula. Many of my colleagues would have added "You idiot. Why didn't you read my 1966 paper? I had this result twelve years ago." Of course as soon as I looked up this reference from Sakharov's smuggled manuscript, I realized that he had had my result long ago, and wrote him a letter apologizing for not referring to his work and for claiming the result as my own original discovery. He never received my letter.

This 1966 paper explains why Sakharov in Gorky should be doing the same work in 1980 that I had been doing in Rehovot. The old 1966 work had been carried as far as possible at that time and there were no new developments for several years. But in the mid 1970's a new theory was developed called Quantum Chromodynamics (QCD) which explained the forces underlying the structure of the particles. The 1966 work made use of the very incomplete knowledge of the forces available at the time. In 1978 it was possible to update the old work by using the new information from QCD. It was natural for me to do this in Rehovot, since I still remembered the old 1966 papers and most of the younger active particle theorists had never known of them. It was also the ideal type of research for Sakharov in isolation in Gorky. He remembered his old work, and is sufficiently brilliant to appreciate the implications of QCD from whatever meager information reaches him. He could combine QCD with his old work without access to scientific libraries or institutes.

Careful reading of the 1966 paper reveals a number of illu-minating observations on the development of physics and on the parallel research carried on in Moscow and Rehovot. A footnote in the paper discusses a formula which Sakharov notes has particular significance. This formula, reproduced in Sakharov's second postcard in Fig. 2, contains a crucial factor of 3/2 which is absent in a similar formula previously derived from an earlier model. The new formula agreed with experiment, whereas the old formula was in strong disagreement. This was presented as additional evidence that the new approach succeeded where the previous approach had failed.

When I looked up Sakharov's reference to the old formula which disagreed with experiment, I was surprised to find a 1965 paper published in *Physical Review Letters* by Haim Harari and Harry Lipkin (!). I had forgotten this work completely, and had to look it up in the library to learn what it was all about. It concerned a point which keeps recurring at different levels in particle physics.

Why are There Always Too Many Particles?

There are always too many particles. Whenever the theorists come up with a beautiful simple explanation for the existing particles, the experimenters upset the system by discovering some new ones which destroy the simplicity and require a new explanation at a deeper level. Theorists are thus always searching for a unification scheme which puts all the particles together in a common framework. The most obvious explanation to a layman is that these tiny particles are made up of even tinier objects arranged in different ways. We have known for a long time now that matter is made up of molecules, molecules are made of atoms, atoms are made of electrons and nuclei, and nuclei are made of neutrons and protons. Why should this stop? Neutrons and protons can be made of smaller objects which can be made of even smaller constituents ad infinitum.

But maybe this does stop. Democritus believed that all matter consists of atoms which are indivisible. The atoms that we know today can be split and divided. They are not Democritus' atoms. But maybe Democritus was right at a deeper level and there are tinier indivisible particles from which all matter is made. This is compatible with another approach to the structure of matter. Instead of looking at the constituents of matter, we can consider the forces between them. Progress in our understanding of these forces has always involved finding new relations between forces which seemed to be very different. Thus the phenomena of electricity, magnetism, radio waves, light and X-rays were found to be all related and described by a single set of mathematical equations for a single field of force, the electromagnetic field. Einstein showed in his general theory of relativity that gravitation could also be described by slightly different field equations. In his later years he searched for a unified field theory which would describe all of the forces of nature in a single set of mathematical equations and would give a unified description of all particles.

Beams of light, x-rays and radio waves were found to consist of particles called photons which are all really the same, although they seem to be very different. Einstein showed in his special theory of relativity that a beam of light on Earth should look like an x-ray beam to an observer in a space ship moving very fast in one direction and that the same beam should look like a radio beam to an observer in a space ship moving in the opposite direction. The photons in light, x-rays and radio waves appear to be very different particles, but are really the same photons seen in different ways.

In recent years theorists have followed two approaches in trying to understand the large number of apparently unrelated particles known at any given time. One approach is to look for new smaller constituents from which everything could be built. The other is to search for Democritus' indivisible constituents and follow the approach of Einstein's attempts to find a unified field theory. The goal is a "grand unification" of all the known forces with all the known particles described as different aspects of the same few particles, like the photons in light, radio waves and x-rays.

In our 1965 paper Harari and I looked for such a "grand unified description" of the particles called mesons and baryons, in the context of an overall symmetry scheme. In this way we derived a relation between the masses of mesons and baryons which disagreed with experiment; our formula looked like the one shown in Fig. 2, but without the crucial factor of 3/2. By 1966 both Sakharov's group in Moscow and our group at the Weizmann Institute had begun to test the idea that all particles were made of even smaller objects called quarks. The 1966 papers from both groups were investigations of this quark model of hadrons, with mesons made of two quarks and baryons made of three. The particular formula noted by Sakharov pinpointed the difference between the grand unification and quark model approaches and showed that one worked where the other had failed. But the quark model was still very controversial and generally ridiculed by the particle physics establishment because quarks had not been observed experimentally and there was no independent evidence that they existed at all. Furthermore there were misleading experimental results like the supposed discovery of a new baryon which could not be made from three quarks. This led Sakharov and Zeldovich to devote the main part of their paper to a possible five quark model, which was discarded almost immediately afterwards.

Today the quark model has become accepted, but there are now too many kinds of quarks. Once again the two approaches are being investigated. The "composite model" approach says that quarks are made of even smaller objects called rishons or preons. The "grand unification" approach looks for a unified field theory that includes everything and in which basic indivisible particles naturally arise and explain all the observed particles. At present the field is completely open. Nobody knows which approach is right, or perhaps they are both wrong. It would be useful if a theoretical formula could be found to distinguish between the two approaches by testing against experimental data. So far no one has found such a formula, like the one of Sakharov and Zeldovich in 1966 which showed a crucial factor of 3/2 that distinguished between the two approaches.

All of this discussion refers to the correct part of the Sakharov–Zeldovich paper which considers the accepted three-quark model for baryons and not the five-quark model they invented to describe the new particle that went away. At the end of their paper there is a note added some time after it was completed, before the proofs went to the printer. This note states that new arguments supporting the accepted three-quark model were presented at a summer school in Balaton, Hungary by R. Socolow from Berkeley, U.S.A. Sakharov did not know that Bob Socolow had come to Hungary after spending a few months at the Weizmann Institute, and that the work he cited in Hungary was done by our group in Rehovot. But it is even more surprising (or perhaps not so surprising) that the basic principle underlying our work was first developed by two young Soviet Jewish physicists in Leningrad. I had learned about this work at an international seminar in Trieste, Italy from one of the Russian theorists from Moscow. He had told it to me almost as a joke, because nobody really took the quarks seriously at that time. The basic point needed by Sakharov was already in this Leningrad work, but he had not yet heard about it and only learned of it later on via Trieste, Rehovot and Hungary.

Further reading of the Sakharov–Zeldovich paper shows more connections with the work at the Weizmann Institute, this time in the part of the paper which turned out to be wrong. Two references are quoted for the discovery of the peculiar new particle with an electric charge three times that of the electron. One is by a group of leading physicists at the University of California; the second is by a group from the Weizmann Institute, analyzing data taken at the European CERN accelerator in Switzerland. There is also a reference to a 1964 paper by Harari and myself, discussing how particles with a charge three times that of the electron could be fit into the old unification scheme. At this confused period in the development of particle physics, the nature of the existing particles was still so unclear that both groups investigated all possibilities.

Elements and Compounds

The reason that the quark model was not accepted easily by the physics community was because nobody has ever succeeded in breaking up any particle into its constituent quarks. We know that water is a compound made of hydrogen and oxygen because we can make water out of hydrogen and oxygen by burning them together, and we can break water up into hydrogen and oxygen by passing an electric current through it. But nobody has broken up a proton into three quarks or built a proton out of quarks.

The chemists called hydrogen and oxygen elements because they were unable to break them up into their constituents by other means. But soon there were very many elements and scientists began to search for some unifying principle. Mendeleev found that they could be arranged in a periodic table which exhibited groups of simply related elements. Then it was found that the elements and their atoms were not indivisible but were made of electrons and nuclei. Experiments were performed which broke up atoms into electrons and nuclei. But soon there were too many nuclei, and it became clear that they could not all be elementary either. Experiments were performed which broke up nuclei into neutrons and protons. At this stage neutrons, protons and electrons were believed to be *the* elementary constituents of matter.

But now something new and unexpected happened. Electrons were also emitted sometimes from nuclei, even though we now know that nuclei contain only neutrons and protons and no electrons. This turned out to be a manifestation of the equivalence of matter and energy, described by Einstein's famous equation $E = mc^2$. Matter can be turned into energy, and energy can also be turned into matter. In this case the energy of the motion of the neutrons and protons in the nucleus could be turned into matter and create new particles like electrons which had not been there before.

The new development of conversion of energy into matter completely upset the further progress of probing the basic constituents of matter. Up to the discovery of the proton and neutron, there was one simple rule. Just hit anything hard enough and it will break up. Physicists attempted to break up the proton by banging two protons together at very high speeds. But the protons did not break up. Instead the enormous energy of the motion of the protons was converted into new matter, and many new particles were created in the collision.

This fit into the general picture that protons, neutrons and electrons were really elementary because they could not be broken up. High energy collisions between electrons produced a shower of photons and more electrons, in agreement with the laws of electrodynamics and the new quantum theory. Similarly, collisions between protons and neutrons produced a shower of mesons which were believed to be analogous to photons in a new field theory which would describe the strong nuclear forces.

But there were soon too many particles. Many new particles like the proton and neutron were discovered and called baryons. Many varieties of mesons were also discovered. They could not all be elementary. The unitary symmetry model of Murray Gell-Mann and Yuval Ne'eman found the analog of Mendeleev's table for mesons and baryons. Everything fit beautifully, and new particles which were obviously missing were searched for and found in experiments, like the missing elements in Mendeleev's table. But the underlying structure was still unclear. Were all the mesons and baryons indivisible, and simply different ways of looking at a few basic particles, like the photons in light, radio waves and x-rays? Or were they compounds, made of tinier objects?

By 1966 the idea that mesons and baryons were made of quarks was being considered, but everyone wondered why quarks had not been seen. They had not only never been seen; they were also believed to have very peculiar properties. The electric charges of the quarks were either two-thirds or one-third of the charge of the electron. No particles with such fractional charges had ever been observed experimentally, and it was generally believed that if such particles existed they would have been seen. Today we believe that the forces holding the quarks together in a proton are so strong that we will never be able to break a proton up into isolated quarks. All the energy we put into trying to break up a proton will simply be dissipated into making more particles. But in 1966 this seemed absurd. Nobody was willing to admit the possible existence of particles which could not be isolated and observed.

How To Test The Quark Theory

The problem formulated both in Moscow and Rehovot was how to test the idea that mesons and baryons were made of quarks, even though individual quarks had not been observed, and their exact properties and the forces between them were not known. The basic idea was very simple. There were three different types of quarks known. Today these types are called "flavors". Quarks of two flavors, called "up" and "down" are found in the proton and neutron. There are also heavier quarks, called "strange" quarks, found in other particles. All the different kinds of particles created in high energy collisions between protons are supposed to be built from these same quarks in their different flavors arranged in various ways. Although the masses of the quarks and the forces between them were unknown, it was assumed that these masses and forces were always the same regardless of how the quarks were arranged in different particles. With this assumption both the group in Rehovot and Sakharov and Zeldovich in Moscow were able to write down formulas which predicted the masses of particles containing different numbers of up, down and strange quarks arranged in different ways. The mass of any particle was simply the sum of the masses of its constituent quarks, plus the mass that came from the energy of the forces between quarks. The experimental values of the masses agreed with these formulas, thus supporting the idea that particles were made of quarks. But there were uncertain factors in the formula because the nature of the forces between quarks was completely unknown.

Another difficulty which perplexed many physicists was an apparent inconsistency between between the quark model and an accepted general principle of physics called "Fermi statistics". This principle states that two identical quarks of the same flavor cannot occupy the same point in space at the same time. But the standard model for the proton did have two quarks of the same flavor at the same point. This bothered Sakharov and Zeldovich, who concluded that the proton could not be made out of three quarks. This gave them another reason for believing in a five quark model for the proton, which was later shown to be wrong. The Rehovot group ignored this statistics problem completely, assuming (correctly as we now know) that it was one aspect of the model which was not understood at the time, and could be left to be solved later on. In fact the solution accepted today had already been proposed by O. W. Greenberg. In Greenberg's model quarks have another property in addition to flavor, which is now called "color". The three quarks in the proton were required to have different colors. Thus the quarks of the same flavor at the same point in space were no longer identical and no longer violated the principle of Fermi statistics.

Since 1966 there have been great advances in our understanding of the forces between quarks as the result of the work of many theoretical and experimental physicists all over the world. The new theory of Quantum Chromodynamics (QCD) is a beautiful generalization of the well-known theory of electricity and magnetism (Quantum Electrodynamics). The strong forces between quarks are described by QCD with a new field of force, analogous to the electromagnetic field. The new field is called the "gluon" field because it provides the "glue" which holds the quarks together in protons. There are two kinds of forces between quarks called "chromoelectric" and "chromomagnetic" directly analogous to ordinary electric and magnetic forces between electrons and nuclei in atoms.

There were also experimental discoveries of new very heavy quarks of two new flavors, called "charm" and "bottom", since 1974. Physicists have now found the quark of a sixth flavor, called "top" which is predicted by the new theories to be the partner of the "bottom" quark.

The QCD theory tells us that quarks of all flavors have the same chromoelectric charge and the same chromoelectric forces, whether they are up or down, strange or charmed, top or bottom, just as the ordinary electric force between two protons is the same as the ordinary electric force between two electrons. But the chromomagnetic force is weaker for the heavier quarks than for the lighter quarks, just as the ordinary magnetic force on a proton is 1000 times weaker than the ordinary magnetic force on the much lighter electron. The chromomagnetic force between two quarks, which behave like tiny magnets, also depends upon the directions in which these magnets are pointing. They can either attract or repel one another.

Updating the 1966 Mass Formula

In 1975 Sakharov extended the application of his old 1966 mass formula to the newly discovered particles containing charmed quarks. In 1980, sitting in Gorky, he improved the model with the aid of the new ideas from QCD. At the same time I did a similar updating of our 1966 work and obtained essentially the same results as Sakharov. The additional information from QCD that the previously unknown interquark forces are chromoelectric and chromomagnetic can be used to obtain new relations between masses of particles. The chromomagnetic force is inversely proportional to the quark masses and depends in a well-defined way on the directions the quark magnets are pointing. The chromoelectric force is the same for all quarks and does not depend upon flavor or direction.

The neutron consists of three light quarks, two downs and one up. A particle called the lambda has one up, one down and one strange quark. It is different from the neutron only by having one of the two down quarks replaced by a strange quark. The lambda is then heavier than the neutron because the strange quark is heavier than the down quark. If this is the only difference, then the lambda must be heavier than the neutron by an amount exactly equal to the difference in mass between the strange quark and the down quark. It is just this relation which appears as the equation in Sakharov's first postcard shown in Fig. 1. The quantity $m_s - m_d$ is just the difference between the mass of the strange quark denoted by s and the down quark denoted by d. The quantity $\Lambda - N$ is the difference between the mass of the lambda denoted by Λ and the mass of the neutron, denoted by N.

The masses of the neutron and the lambda have been measured very precisely in laboratory experiments. The masses of the quarks have never been measured at all, because nobody has been able to find isolated quarks and do experiments with them. But the equation in Sakharov's postcard tells how to obtain the value of the mass difference between the s and d quarks from the measured values of the masses of the neutron and lambda.

We can now examine other pairs of particles like the neutron and lambda which differ only by having a down quark replaced by a strange quark. We can play the same game with the particles called mesons which are made of two quarks. Some mesons are made of one up quark and one down quark; some heavier "strange" mesons are made of one up quark and one strange quark. The strange mesons are heavier than the light mesons also by just this mass difference between the strange quark and the down quark. The masses of these mesons are known from experiments in the laboratory. The mass difference between the quarks is now known from the equation in Sakharov's postcard and the known values of the masses of the lambda and neutron.

We now have a theoretical result that can be checked against experiment. If this model is correct, the experimental values for the masses of the strange mesons must be heavier than the masses measured for the light mesons by exactly the same amount as the lambda is heavier than the neutron. This result is also stated in Sakharov's first postcard in Fig. 1. This relation between the lambda–neutron mass difference and the mass difference between strange and nonstrange mesons agrees very well with the experimentally measured masses of these particles. I was very excited when I found this result in 1978, without knowing that Sakharov and Zeldovich had already found it in 1966, because it confirmed the hypothesis that mesons and protons are made of the same quarks arranged in different ways. Similar relations can also be obtained using charmed quarks and quarks of other flavors.

But life is not quite that simple. There is one subtle point in this relation. The difference in mass between a lambda and a nucleon is not only the difference in the mass between the strange and down quarks. There are also the energies of the forces between quarks which must be properly taken into account. Before QCD there was no theoretical basis for computing the energies of these forces. This is why the Rehovot group could not obtain this result in 1966.

Sakharov and Zeldovich made an inspired guess that anticipated QCD by assuming that the forces between quarks were similar in behavior to the forces between electrons in atoms. In this way they were able to obtain the result shown in the postcard. They assumed two kinds of forces analogous to electric and magnetic forces, that the electric forces were the same for quarks of all flavors, and that the magnetic forces depended on the direction of the quark magnets like the forces between ordinary magnets. This gave the result which we now know from QCD that the chromoelectric forces are the same in the neutron and the lambda and do not affect the mass difference between them, and similarly for the mesons. The exact values of the chromomagnetic energies cannot be properly computed without QCD. But the assumptions of Sakharov and Zeldovich were sufficient to enable them to chose the right combinations of particles so that the contributions of the chromomagnetic energies exactly canceled out. This is the reason for the numbers (1/4) and 3 in the postcard of Fig. 1. The K and K^* mesons have one strange quark; the pi (π) and rho (ρ) mesons have no strange quarks. But all these mesons have different chromomagnetic energies. The particular combinations chosen cancel out the chromomagnetic energy. This is the reason for Sakharov's remark that "of course you are right and $m_s - m_d = \ldots$ and not the decuplet mass splitting." The "decuplet mass splitting" was the old-fashioned estimate of the quark mass difference, which did not take proper account of the energy of the forces between quarks.

Additional relations between particle masses are obtained by looking at two particles believed to be made of exactly the same quarks and differing only by having their "chromomagnets" pointing in different directions. The masses of these two particles differ only by the change in the energy of the chromomagnetic force. We can now compare the chromomagnetic energies for such particle pairs made of different combinations of the light up and down quarks, the heavier strange quarks and the very heavy charmed quarks. The QCD result that the chromomagnetic energies are inversely proportional to the masses of the quarks can be used to obtain relations between these mass differences. The values of the experimentally measured masses are in excellent agreement with these relations.

How Particles Behave Like Tiny Magnets

Another test of the quark structure of mesons and baryons is provided by the measurements of the behavior of these particles in magnetic fields. It has been known for a long time that protons, neutrons and nuclei behave like tiny magnets when they are placed in magnetic fields. The nucleus is a magnet made up of many tinier proton and neutron magnets. The strength of the magnet of the whole nucleus depends on how the magnets of the individual protons and neutrons line up with one another. Thus precise measurments of the strengths of these magnets gave very useful information about how the protons and neutrons fit together into nuclei.

In the same way the neutron and proton magnets are made of tinier quark magnets and the strengths of these magnets depend upon how these quark magnets line up with one another. The quark model for the neutron and proton tells exactly how they line up. The prediction that the neutron magnet should have exactly two thirds the strength of the proton magnet is in striking agreement with the experimental values which had been known for a long time but not understood theoretically. This was one of the first successful tests of the quark hypothesis back in 1964.

It was natural to extend these ideas to predict the strengths of the magnets of the other particles made from three quarks, like the lambda. This was in fact done at the Weizmann Institute in 1966 by the group of Hector Rubinstein, Florian Scheck (one of the first Minerva Fellows from Germany), and the same visitor Bob Socolow whose report on our work later in Hungary reached Sakharov and was acknowledged in their paper.

But there were two difficulties, one theoretical and one experimental. The theoretical problem is that the lambda contains a strange quark, and the strength of its magnet depended upon its mass, which was not known. Thus the strength of the lambda magnet could not be directly predicted from the known proton and neutron magnets.

The experimental problem was that the strength of the lambda magnet was exceedingly difficult to measure in the laboratory and not very well known at that time. Protons and neutrons can be produced in large numbers in the laboratory and kept in an experimental apparatus for a long time while magnetic measurements are made. But the lambda particle can only be produced in very high energy accelerators by banging two protons together, and it lives for only a tiny fraction of a millionth of a second before it breaks up into other particles. The magnetic measurements on the lambda can be made only by producing them at an accelerator and passing them through a magnetic field during their tiny lifetime. Thus it was not until the summer of 1978 that a precise measurement of the lambda magnet was available.

The formula in Sakharov's postcard gives the mass of the strange quark and can be used to predict the strength of the magnet of the strange quark and of the lambda particle. But neither Sakharov and Zeldovich in 1966 nor I in 1977 thought of using these results to calculate the magnetic properties of the lambda. However when I arrived at the Fermi National Accelerator Laboratory (Fermilab) near Chicago in the summer of 1978, I learned that a group of high energy physicists from Fermilab and the Universities of Michigan, Rutgers and Wisconsin had suceeded in producing a beam of lambda particles and measuring the strength of the lambda magnet very precisely. When they told me their result, which was not yet published, I was very excited, because it fit exactly with the value of the mass of the lambda that I had obtained earlier. I wrote a paper on it and had to wait before submitting it for publication in *Physical Review Letters* until the experimental group had officially announced their value for the lambda magnet.

Andrei Sakharov in his isolation in Gorky had no idea that the strength of the lambda magnet could be measured so precisely and that a value was about to be published. Had he been free to visit the great laboratories of the world and to hear about their latest exciting results, he would certainly have made even greater contributions to the investigations of particle structure. It is remarkable that he has managed to produce the original research that he has under his difficult conditions.

Now, although nobody has succeeded in breaking up the proton into quarks and showing that isolated quarks exist, there are many such experimental tests of the theory that particles are made of quarks bound together by chromoelectric and chromomagnetic forces, and that the strengths of these forces depend upon the masses of the quarks and the directions of their magnets in the way suggested by the QCD theory. But the last word on QCD has not yet been said. There are still many more experiments which must be performed using higher and higher energies to see whether the particles really behave in the manner predicted by QCD. And there are many complicated theoretical calculations which must be made in order to determine what QCD predicts for these experiments. QCD calculations are extremely complicated. There is as yet no recipe for calculating the result of an experiment from the equations of QCD like the recipes for calculating the orbits of planets, satellites and rockets from the equations of motion of Newton and Einstein. Theorists are working hard to find ways to cope with the complicated equations of QCD, but still have a long way to go. There is still much exciting work to be done.

The story of elements compounds and forces continues. Are the quarks and the leptons like the electron really elementary, or are they compounds made of even smaller building blocks? If they are compounds, what are these new building blocks, and what kinds of new forces hold them together? These are the same questions that Sakharov and Zeldovich asked in Moscow in 1966 and the Weizmann Institute group asked in Rehovot about the proton and neutron. Now these are answered and we are back at the same question at a deeper level. Thus the search for new knowledge at the frontier goes on.

4

How Scientists Study Nature — Pure and Applied Research

What is all this research on the structure of matter good for? Who needs it? To understand this we must understand what science is, the difference between real science and pseudoscience and the gap between natural and social science.

In 1946 graduate students at Princeton were concerned about the effects of the Atom Bomb and the need to take future development away from the military. A group went to Robert Oppenheimer, then head of the Princeton Institute of Advanced Study, for advice on what to do to be effective. Oppenheimer told us that the gap between natural science and social science was too wide. We should organize a series of lectures alternating between top scientists in both areas. We followed his advice and the talks turned out to be very interesting. As a result of these talks Bob Bush, the main organizer of our group, left experimental physics and took an interdisciplinary fellowship at Harvard. He later became head of the Psychology Department at Columbia University.

In his first year at Harvard, Bob came back to visit Princeton and told us to go into social science. These people have no idea of what an experiment means. We physicists were astounded at their distorted view of the "scientific method" presented in courses in sociology.

You take a hypothesis and test it in experiments. There is no indication of where the hypothesis came from. No indication of how one chooses a hypothesis to test. No indication of what to do with the results of the test. They learn nothing from experience.

A similar point of view has been expressed by Nobel laureate Richard Feynman, discussing "The pleasure of finding things out", with some very harsh comments about the social sciences and their methodology and the gap between natural and social science. "Because of the success of science, there is, I think, a kind of pseudoscience. Social science is an example of a science which is not a science; they don't do things scientifically; they follow the forms — you gather data, you do so-and-so and so forth but they don't get any laws, they haven't found out anything. To know how hard it is to get to really know something, how careful you have to be about checking the experiment, how easy it is to make mistakes and fool yourself. I know what it means to know something, and therefore I see how social scientists get their information and I can't believe they know it, they haven't done the work necessary, haven't done the checks necessary, haven't done the care necessary. I have a great suspicion that they don't know, that this stuff is wrong, and they're intimidating people."

Natural scientists choose a topic for investigation, see how this fits in with present knowledge and try to find ways to create new knowledge. They then use the new knowledge to find the next topics for investigation and apply this new knowledge to important real problems.

The twentieth century began with the consensus that matter is not continuous but is made up of atoms and molecules. It ended with the confirmation that matter is made of even tinier objects called quarks. But how are such realities established? And how can the process be explained in a plausible way to a non-scientist?

I was surprised to find the answer in a book, *The Schools We Need and Why We Don't Have Them* (Doubleday, New York, 1996), by E. D. Hirsch Jr, a professor of English at the University of Virginia.

I read this book because of my concern about the large numbers of functionally illiterate children in both Israel and the United States unnecessarily condemned to the lowest levels of a democratic society. Professionals in the field of reading education are still obsessed with fighting wars between outdated "ideologically and politically correct" party lines called phonics and whole language, rather than using new knowledge obtained from research and classroom experience to teach children to read.

How to use this knowledge to provide effective education is well documented in extensive research sponsored by the U.S. National Institutes of Health. Twenty years of experience in Israel have shown how all this knowledge and information can be combined to produce a system that can teach 95% of a heterogeneous class of as many as 40 six-year-old pupils to read, to understand written texts, and to become independent learners with a single teacher and no help from parents.

Not only does Hirsch's book provide the theoretical background supporting the empirical success of this system, he also gives an excellent description of how the scientific community reaches conclusions. He writes: "The pattern of independent convergence (a kind of intellectual triangulation) is, along with accurate prediction, one of the most powerful confidence-building patterns in scientific research. There are few or no examples in the history of science (none that I know of) when the same result, reached by three or more truly independent means, has been overturned." He quotes Abraham Pais's biography of Einstein for an example of this convergence: "The debate on molecular reality was settled once and for all because of the extraordinary agreement in the values of N (Avogadros number) obtained by many different methods. Matters were clinched not by a determination of N but by an overdetermination of N. From subjects as diverse as radioactivity, Brownian motion, and the blue in the sky, it was possible to state, by 1909, that a dozen independent ways of measuring N yielded results which lay between 6.21×10^{23} and 9.21×10^{23}."

In 1966, Richard Dalitz and I were both already convinced that matter was made of quarks when we led the discussion on this topic at the annual international conference on high-energy physics. We could not understand why this conclusion was not generally accepted until well into the 1970s. I now realize that we were experiencing Hirsch's independent convergence. Unexplained regularities in the spectrum of particles created in high-energy accelerators; simple relations between collisions among different kinds of particles; that the annihilation of a proton and an antiproton at rest nearly always produced three mesons; relations between the electromagnetic properties of mesons and baryons; and the 3/2 ratio of the magnetic moments of the neutron and proton all converged on the same conclusion: mesons and baryons are built from the same elementary building blocks.

Hirsch applies this concept to pedagogy: "The independent convergence on the fundamentals of effective pedagogy that exists today is less mathematical but nonetheless compelling." I find Hirsch's demonstration of how the scientific community reaches conclusions and of how the results of research on reading can be interpreted very exciting. I have always disliked the explanations of the "scientific method" presented by social scientists in which scientists are said to do experiments to "test hypotheses". The description of the acceptance of new ideas as a result of independent convergence fits the reality that I know.

The gap between natural and social scientists is much too large and deserves serious attention. Perhaps these ideas can stimulate new lines of communication to overcome this gap. Perhaps they can also direct us to using new knowledge in finding better ways to teach children to read.

Like science, teaching should be the result of independent ideas converging. As an example of how scientific research can help society solve real problems we consider reading education. As a physicist my scientific approach different from the pseudoscientific approaches of the reading experts who are in all the Universities at Departments of

Education. I can understand the importance of the new knowledge obtained from scientific research about reading education.

Unfortunately scientific research has been replaced by politics in most university departments of education. There reading wars between two political reading approaches emphasizing sound vs. meaning have dominated conventional thinking and led to pendulum swings in which nothing new is learned at each transition.

A scientific approach bypasses these by recognizing that every first grade pupil already knows that words have both sound and meaning. Scientific knowledge pinpoints what is missing and crucial in both approaches. Enormous new knowledge about how children learn to read has accumulated over the past half century from scientific research on linguistics, cognitive psychology, brain research, information processing and more.

A child entering first grade with the ability to speak and understand a normal first-grader's vocabulary has in his brain a memory and an information processing system far superior to anything that Silicon Valley will be able to supply for a long time. And he/she is every day developing new hardware, new software, new memory capacity, new interconnections, and maybe even all kinds of new things still unknown to Silicon Valley. This enormous amount of information already stored in a child's brain can be used to help him learn to read. The basic problem faced in reading education was clearly stated by the distinguished linguist Alvin Liberman:

"Why is Speech so much Easier than Reading and Writing? What more does a child who already speaks and understands his native language need to learn in order to be able to read?"

New scientific knowledge has taught us that humans are born with the ability to use two codes, the genetic code and the language code. Each code uses a small number of basic building blocks each having no intrinsic meaning. Combining these blocks in different ways produces an enormous number of meaningful messages to control the senses and muscles of the body. In the language code the basic blocks of most written languages is the alphabet. But preliterate

51

humans used the language code before the invention of the alphabet. Information technology and computer science tell us that a child who has already learned how to speak and understand speech has a tremendous amount of information already stored in the memory of his/her brain. This information is stored in a way that enables instantaneous access of any word and its meaning whenever it is needed for speech or for understanding the speech of others.

That words have both sound and meaning is understood by every child who knows how to speak and understand the speech of others. His brain has already stored the incredible amount of information needed to produce speech by activating the nervous system and the appropriate muscles in the lips, tongue, vocal chords, larynx, etc. as well as the information needed to process the sounds he hears and turn them into words and meaning. All this is already stored in a child's memory by the time he has learned to speak. The remarkable, almost miraculous nature of this achievement is easily appreciated by anyone who has tried to store information into the memory of a computer and get it back later. So I am quite prepared to accept the statement by experts that some form of code is involved in this process of information storage and recall.

There is much talk about emphasizing "practical" applied research rather than pure "ivory tower" research. Students and young researchers should be encouraged to pursue useful directions of research which will lead to the solutions of the important technical problems facing society and to the development of new inventions which will improve the quality of life. But how is this done? Who decides which are the right directions?

Some good examples of the problems involved in directing students and in understanding what research is all about and what it is good for are illustrated in my own personal experiences. In 1938, I entered the Electrical Engineering School at Cornell University and wanted to study something practical. The electrical engineering curriculum then included only one semester of electronics, and the professors refused to listen to requests from the students for more electronics. They assured us that although we liked to play around with ham radio, electronics was not practical and there was no future in it. There were no jobs in electronics. Ninety per cent of their graduates found jobs in power engineering, and we had better study our machinery and power transmission courses and forget about this useless electronics.

Had we been students in the 1960's, we might have organized protests and demonstrations to change the curriculum. But in the 1930's the attitude was different. Fortunately the university was flexible and allowed students to take additional courses of their own choosing. Some of us took extra courses in physics and mathematics. We had heard that there were very interesting courses given over in the physics department by two new refugee professors from Hitler's Europe, Hans Bethe and Bruno Rossi. So we went over to the physics building and listened to them.

In the engineering school we learned that electrical energy traveled through wires. The engineers also knew that radio existed and that electrical energy also traveled through the air. But they didn't really understand it and it wasn't practical. In the physics department we learned about the basic properties of matter and energy without any pretense that this was practical. We also learned how electrical energy traveled through the air, as described by the famous equations of Maxwell. Engineering students did not study Maxwell's equations in those days.

When I graduated in 1942, the U.S. had entered World War II, and I joined the massive effort at the Radiation Laboratory at M.I.T. in the development of microwave radar. This very important research was all based on electronics and electrical energy traveling through the air, which were both considered impractical only a short time before. My work began with a three month course in microwave and pulsed electronics, which we had not learned in our university training, because electronics was not "practical". This taught me an important guideline for work in science and technology. What is most practical today will probably be out of date tomorrow.

The microwave radar program was an outstanding success. One of its major achievements was detecting German submarines from the air. This was possible because the German establishment had made a high level decision that radar at microwave frequencies was not practical, and their submarines were not equipped with microwave receivers which would warn them of an approaching airplane carrying microwave radar.

I was surprised to find that the major contributions at the Radiation Laboratory were made by physicists rather than electrical engineers. Engineers with the standard training, aimed at specializing in the "practical" directions were unable to cope with the new phenomena of high frequencies and wave guides. The key people at the lab were all physicists, not engineers and the staff included half a dozen physicists who later won Nobel prizes. The engineers had learned how to solve the problems that were practical today, but were unprepared for the completely new problems that become practical and even urgent tomorrow. They knew all about how electrical energy traveled in wires but could not understand how it could travel through the air. They did not know how to design radar equipment and make it work.

A radar receiver that I developed was produced by a well known industrial company in Chicago. They built the first model according to my original design, but their engineers introduced small changes to facilitate mass production. It didn't work, because these changes allowed electrical energy to travel through the air in peculiar ways that completely ruined the performance of the receiver. They did not realize this of course and blamed everything on some mistake in my original design. I had to make a special trip from Boston to Chicago to tell them to how to get the receiver to work. After examining their model I told them to move a wire soldered at one point on the chassis to another point on the chassis a centimeter away. They looked at me as if I were crazy. Every electrical engineer knew that when a chassis was grounded it made no difference where you soldered a wire to it. They had never worked with such high gain high frequency amplifiers before. They moved the connection to humor this young fool and were amazed when all their troubles went away and the receiver worked. To them it seemed like black magic.

Shortly after the war ended when I was making plans for what to do after leaving M.I.T., an outstanding scientist from Palestine, Prof. Ernst Bergmann, visited Boston. I asked his advice on how to best prepare for eventual immigration to Israel, and what sort of experience or education would be useful. At the time I considered three possibilities: geophysical prospecting for an oil company, a computer project which later turned out to be one of the pioneering attempts at modern electronic computers, and graduate study in pure physics. Prof. Bergmann said that geophysical prospecting and computers were not important for Palestine and that graduate work in pure physics was probably the best choice.

After working four years at M.I.T. as an electronic engineer, I decided to go to graduate school in physics and study the basic properties of matter and energy, rather than more "practical" subjects, in order to be better prepared for future developments. At Princeton I also learned the exotic games played by the mathematics and physics graduate students every afternoon at tea time, including a variation of chess called kriegspiel (war game). In ordinary chess each player sees all the opponent's pieces and knows his moves immediately. This was considered uninteresting by the students. In war the other side's position and movements are completely unknown, except when there is contact, and intelligence tactics are an integral part of any strategy. Kriegspiel introduces this war-like situation by having the two players seated back to back, each with his own board and pieces. Each moves in turn according to the rules of conventional chess, but cannot see his opponent's board and does not know his moves. A referee watches both boards to ensure that everything is legal. The referee informs the players whenever a piece is captured, whenever a king is in check and whenever a particular move is illegal because it is blocked by an opposing piece or because it is moving into check. But this is the only information they receive and they have to guess at their opponent's position from these clues. This differs from conventional blindfold chess where the player is told his opponent's moves and simply has to remember them. A standard tactic of the game is to try suspected illegal moves in order to obtain information about the locations of opposing pieces.

Kriegspiel is very amusing for the spectators, who see both sides. As each player obtains information about his opponent's pieces, he moves them around on his board to the places where he thinks they should be. Often his picture is very different from the reality seen by the spectators. Good players keep checking their picture of the opponent's position by trying moves which they know to be illegal if their picture is correct. Often these checks happen to give the expected answer, just by chance, even when the picture is completely wrong. The spectators eagerly wait to see how long it takes before a crucial test shows that the theory is wrong.

Kriegspiel is good preparation for frontier research. The science that we learn at the University is like chess. Everything is clear and known. But at the frontier the scientist is playing Kriegspiel against nature. He has to find out what is lurking there in the unknown, with the limited means of testing and exploration available. He must construct crude theories and models on the basis of incomplete information and continually test them and refine them. Often his explorations lead to an apparently confirmed theory, which later turns out to be wrong. Columbus' search for a short route to India found land populated by people who were erroneously called Indians and the misnomer survives until today. Exploring scientists make mistakes like this all the time, and they have to be sorted out.

All this makes one wonder about how to direct promising young scientists toward fruitful applied work. My professors at the university could not foresee the importance of electronics. Professor Bergmann could not foresee the importance of geophysical prospecting and computers in Israel. But Prof. Bergmann's advice to obtain a good background in physics was really the best.

Graduate study in a good university in a pure science provides the training necessary for work in new areas which cannot possibly be anticipated at the time the student begins his studies. The student learns to solve new problems by developing new techniques and discovering new things. Exactly what he develops and what he discovers at this stage is not so important. It is learning the approach to search and discovery and gaining experience in attacking new problems, where one cannot find the techniques for solution in any textbook or handbook, and one has to work it out all alone.

What is most practical today will probably be out of date tomorrow. But it is not so simple to say that researchers must concentrate on directions which will be important to the country in the future. Who can predict the future? When I hear some people trying to tell other people what they should be doing, I am reminded of the words my father used to say to me when I thought I had been very clever. "If you knew what you don't know, you would know more than you know."

This is the basis of all learning. It doesn't matter whether you are a child learning to read, a biomedical scientist searching for a better treatment for Parkinson's disease, a computer scientist trying to improve a word-processing program or an astrophysicist looking for the keys to the cosmos. Everyone is working at the frontier between the known and the unknown, trying to push the frontier a little further and know more today than he did yesterday. Each push gives satisfaction, enjoyment and a feeling that what has been learned may be useful. But no matter how far we go, there will always be much more to be learned than what we already know.

We continue the example of reading education by showing how the results of scientific research are applied.

The task of a reading teacher is to know where each individual pupil is at the pupil's own frontier between known and unknown and to provide the guidance and tools enabling the pupil to know more today than yesterday and be prepared to know even more tomorrow.

And the excitement of being at the frontier and pushing beyond by discovering new knowledge is always maintained along with progress toward the goal of reading mastery.

What we don't know about child development is still much more than we know. But what we already know seems to be enough to create a successful system for teaching reading.

A child who speaks and understands spoken language is at the frontier between knowing how to speak and understand and learning how to read and write. The question arises how can this enormous amount of information that is already stored in a child's brain be used to help the child learn to read. The language already known in a spoken language code needs to be converted into a reading language code whose basic building blocks are the alphabet. The child already knows how to use the language code without being aware of its basic sound building blocks. The goal of a reading teacher is to enable the pupil to be aware of the basic sound building blocks. This is called phonemic awareness. Once the child can isolate the basic building blocks in sound the next step is to learn how to represent these basic building blocks by an alphabet.

The phonetic alphabet was invented by the Phoenicians who created a Semitic language alphabet similar to that used in Hebrew. This alphabet represents each letter by a single sound. There is no letter like the letter "c" in English which has two completely different sounds and a third different sound in the combination "ch" which represents a single sound different from the two different sounds represented by the letters "c" and "h". For simplicity I continue this analysis using the Hebrew language.

A preliterate Israeli child who utters the Hebrew word shalom and understands what he is saying is not aware that the word shalom he is saying can be broken up into syllables, vowels and consonants. Similarly a preliterate child who hears the word shalom and understands what he is hearing is not aware that the word shalom he is hearing can be broken up into syllables, vowels and consonants. Preliterate humans developed the ability to communicate using such words for over 200,000 years without discovering the alphabetic principle that enabled them to break up the words they knew into syllables, vowels and consonants and represent them in a written alphabet. The problem was not representing visually what was known to the ear. It was understanding the structure of what the ear recognized instinctively.

Today teaching children to read the information that they know how to communicate by speech means teaching them to learn what it took their ancestors over 200,000 years to discover. They must learn how to break down the words that they have instinctively learned to speak, hear and understand into components that can then be systematically written. The problem in teaching reading is to understand and use the nature and structure of speech that has been instinctively incorporated into the human brain, and to be aware of how and why it works so remarkably well. The problem is not how to transform the audio into video.

To enable a child to read and represent by visual symbols what he/she already knows by speech and hearing he/she must first learn to access his/her own memory and become aware of these basic building blocks that are already there in the memory. The next step is to represent them visually by the alphabetic principle.

The alphabetic principle is the key to reading. But experience has shown that some children begin reading at age three while others are not ready to master the alphabetic principle until they are close to age seven. How can one teach a class which is a heterogeneous mix with pupils at all levels of development?

This leads to the other crucial aspect of reading education. The children are all different. At the first grade level the brains of each are developing new hardware, new software, new memory capacity, new interconnections, all at different levels and different rates. The reading teacher must know where each pupil is and what is needed to provide the guidance needed for progress on a road to the discovery of the alphabetic principle. But realistic budgets and manpower cannot provide an individual teacher for each pupil.

A successful reading education system must give the teachers the training, materials and infrastructure needed to cope with existing heterogeneous classes. Otherwise the teacher is usually forced to concentrate on the middle of the class, boring the brighter pupils and losing the slower ones.

To learn more about how children learn to read I have read articles about dyslexic brains, phonological awareness and processing phonemes in the *Scientific American,* reporting results of brain research using the latest available technologies like MRI. I attended a one-day symposium on Literacy and the Language Module in October 1997, a conference in honor of Alvin Liberman's APA Lifetime Achievement Award, which reported on much new knowledge on teaching reading, human language and brain research obtained at Yale University and Haskins Laboratories.

Teaching children to read must start with using information already stored in the memory of the student's brain. It can then move on to teach the necessary additional information to enable the student to master reading.

How new input from scientific research has been used to help real children in the classroom can be seen in my millennium essay in *Nature,* Vol 406, 13 July 1, 2000, showing a picture of a class using a scientifically based system for teaching reading in Israel. The caption is "95% of six-year-old Israelis can read, understand and learn independently". The system must start with a clear message to beginning first grade teachers: "The children are all different, but we have to teach all of them and we can do it!"

Results show that it works in the classroom by using the enormous new knowledge from scientific research. In contrast to other systems which immediately flood the pupil with new unrelated information like an alphabet, one starts with what is already stored in the memory and leads eventually to the alphabetic principle and mastery of reading, fluency and understanding.

To bypass the political reading wars between sound and meaning one starts with thirty simple words in the pupil's commonly used vocabulary. These "memory-support" words include all the vowels and consonants and remind the pupil when necessary of the sound of a letter. In contrast to other systems which immediately flood the pupil with new unrelated information like an alphabet, one starts with what is already stored in the memory and goes on from there. It leads eventually to the alphabetic principle and mastery of reading, fluency and understanding. Only the sounds of the letters are used at this stage, not their names. The extra effort required to remember the names of the letters is deferred until after the completion of reading mastery. Each pupil is then led on an individual voyage of discovery in seven stages of how words can be broken down into syllables and eventually into their constituent vowels and consonants, while at the same time learning new words and combining them into sentences and paragraphs.

A child who knows how to say the Hebrew word shalom has the information already stored in the memory that the word contains two parts sha and lom, and that sha contains a "sh" and an "a" and that the lom contains an "l", an "o" and an "m". This information enables the child's brain to send the complicated messages through the nervous system to the relevant muscles to produce the elementary sounds and blend them into the word. The code is the information that the same "sh" sound appears in many other words, and that all of the words the child knows are built from a small number of building blocks. The child is not aware of all this. He/she just utters the whole word when needed. The teacher must guide the child to access his/her own memory and become aware of these basic building blocks that are already there in the memory, and represent them visually by the alphabetic principle.

One method for providing this education answer has been given by the LITAF program developed by educator Nira Altalef for teaching reading in Israel by using her own years of classroom experience. As an educational advisor and consultant in junior high schools she learned how children learn and teachers teach. She did not know any of this research nor information technology. We have independently discovered that the LITAF system is using exactly the right procedure to make optimum use of the information already stored in the child's memory when he enters first grade. Twenty years of experience have shown that it works. LITAF gives a teacher the tools to know which pupil in a class of 40 is ready for what, guide his or her development through seven stages on the way to reading mastery, and end with over 90% having mastered reading by the end of first grade, bright pupils continually stimulated with work at their own levels, slower students continually progressing and feeling successful at their own rates, and all having become independent learners and knowing how to learn by themselves. The exciting answer is that it IS being used to teach children to read and that it WORKS!

The key is always the child and not buzz-words and labels like "gifted", "learning impaired", "slow" and "dyslexic". The child is where he/she is in the process of development of learning to read, and it is the teacher's task to know where and how to help the development. Above all, there is never failure. The word "failure" does not exist in LITAF. A child is where he/she is and the fact that he/she may not be where some educator or bureaucrat thinks he/she ought to be is not failure. It is part of the reality of human experience, and is by definition always good. This reality can always be made better and it is the task of teachers, schools and educational systems to find the way how to make it better for everyone at his/her own stage of development. For reading, making it better means finding the ways to enable over 90% of a normal class to achieve reading mastery, comprehension, fluency and independent learning by the end of first grade. LITAF does this.

What is Scientific Research?

These words "If you would know what you don't know you would know more than you know" are a guide to the status of scientific research. No matter how hard the scientists of the world work to accumulate new knowledge, and no matter how much progress they make in understanding the universe, the unknown regions awaiting discovery will always be larger than what is known. The work of scientists today only pushes the frontier between the known and the unknown a little further into the unknown. The most exciting and dramatic breakthroughs by the great geniuses are only tiny steps on the road to new knowledge. The basic question facing every creative scientist is which particular tiny portion of this vast unknown region to choose for his investigations.

There is an old story about a drunk looking under a street lamp for a ring he had lost. A friendly policeman comes to help and asks where he lost it. "Over there about two hundred yards," answers the drunk. "Then why are you looking here?" asks the policeman. "It's dark over there, you can't see anything. Here there is light," answers the drunk.

This drunk characterizes the scientist who explores the frontier region between the known and the unknown, where there is still enough light to guide his search into the unknown. He does not know what he will find, but he hopes that it will be interesting new knowledge which will be of value to mankind.

As a contrast to this example, we might consider a drunk looking in complete darkness for a ring which he has never seen, but which he knows is there. "How do you know that such a ring exists at all?" asks the policeman. "How do you know that it does not exist?" answers the drunk. "Can you prove to me that it does not exist?" This drunk characterizes the pseudoscientific fanatics who are looking for phenomena like perpetual motion machines. They know what they are looking for. They believe that they have some special guidance which directs them to the right point in the vast unknown where they will find the object of their search. They think that they know what they will find. They do not realize that if they knew what they would find they would know more than they know.

Between these two extremes is the drunk standing on the edge of the illuminated area, looking into the twilight zone between light and darkness. He also believes that he is looking for a ring, but he is ready to examine all the new and interesting things he may find on the way. In the end these new things turn out to be more important than the original object of the search, the ring which probably never existed at all. The basis of modern chemistry was laid by the alchemists in their search for a means to turn base metals into gold. The basis for the great advances in physics by Galileo and Newton was laid by the ancient astronomers who made precise observations of the stars because they thought it was the key to predicting the future. Christopher Columbus discovered America by searching for a short route to India. These people contributed greatly to human knowledge because their search was open and not single-minded. They did not find what they were originally looking for, but uncovered new things even more interesting.

This open search contrasts sharply with the work of the single-minded fanatic who is sure that he can construct a perpetual motion machine. He builds one model after another. As each one fails and he understand the reason for its failure, he designs the next one which avoids the difficulty, only to find that the new one has a new and different trouble. At the end of his career he has discovered only that a whole host of machines will not work. But he has added nothing to the sum total of new knowledge beyond the fact that these particular machines, built in the way that he has built them, will not produce perpetual motion.

How does a physicist decide which directions of research are promising? Many laymen are misled by the picture of the great genius sitting at his desk, contemplating the mysteries of the universe, suddenly getting a great idea, developing a theory and testing the theory by performing experiments. The direction of scientific research does not come from revelations to great men sitting in ivory towers. It comes from experiments performed at the boundary between the known and the unknown. When these experiments present puzzles and paradoxes which cannot be understood on the basis of previous knowledge, then physical research becomes exciting. It is then that the great theorists try to find the way to bring together apparently incompatible phenomena with a new approach.

Einstein's theory of special relativity followed a series of perplexing experiments involving the relation of light and moving bodies. Experiments performed with light from moving sources, light observed by moving observers and light passing through moving media gave peculiar results which could not be understood on the basis of previous concepts used in physics. Einstein showed that drastically revising the basic concepts of space and time led to a consistent description of all these perplexing experiments and also predicted many new interesting effects which were subsequently confirmed by experiment. These new concepts are accepted today because they successfully described puzzling experiments and led to new fruitful predictions. They are not justified today by theoretical contemplation about space and time.

An active research scientist today must always keep in mind the Principle of Humility. He must constantly say to himself "Anything I can do Leonardo da Vinci could do better — or Einstein, or Newton, or Galileo, or Archimedes, or the great anonymous geniuses who found the way to make bronze from copper and tin in the bronze age, invented the wheel, developed the skills to build pyramids and invented agriculture." For thousands of years, thousands of great men have been making great discoveries. Today's research scientists are trying to find new things which these great men were unable to find. How can they hope to succeed where better men than they have failed? There is only one way. Today we have new tools and new knowledge which were not available to those great men of past generations, and which are available today as a result of all of their work. Only by using these tools and knowledge which they did not have can we hope to find what they did not find.

Scientific progress results from many millions of little discoveries made by different people. Each new work builds on previous discoveries. The great breakthroughs come when many new discoveries accumulate and reach a point where they can be brought together and used to lead to a deeper understanding and further new discoveries. Galileo invented the telescope and used it to discover the moons of the planet Jupiter. Without the telescope the moons could not be discovered; they are not visible to the naked eye. But the telescope could not be built without good quality optical lenses. The technology of lenses was developed for eye glasses, but the lenses used for eye glasses were not sufficient to make a telescope. It was only after lens technology developed to the point where many good quality lenses were easily made, including those which were completely useless for eye glasses but suitable ingredients for telescopes, that telescopes became feasible. Once the lenses were available, it was only a matter of time before someone would think of putting a few of them together and discover the principle of the telescope. But before the lenses were available it was impossible.

The moons of Jupiter were there all the time, waiting to be discovered. But the lenses needed were not developed by people who wanted to discover Jupiter's moons. They were developed by people looking for better and more efficient ways of making lenses for eye glasses. Galileo did not invent the telescope in order to look for Jupiter's moons. But once he had it, it was natural to use this new tool to explore new horizons. Many great scientific discoveries have occurred in this way, by developing the necessary tools without any intention of making this particular discovery. After a new tool is available it is used to explore the unknown wherever possible to find new things.

In modern science, breakthroughs have occurred very soon after the necessary tools were available. The tools did not sit around for a century waiting for a great genius to come and show mankind how to use them. There have been no great breakthroughs which were possible a hundred years earlier. If the breakthrough was possible, somebody found it. Great discoveries in science often occur simultaneously in different places. When the time comes, and all the knowledge needed for a breakthrough is available, several great scientists may see it independently.

The transistors, lasers and computers of today were impossible forty years ago. There was not enough basic knowledge and technology available for their development. It would have been like trying to make telescopes before the development of adequate lenses. They were developed only after the necesary basic knowledge was provided over a period of years by research into the properties of materials and the interpretation of experiments with the new quantum theory of the atom. Physicists investigating the way electricity passes through different substances found peculiar materials called semiconductors which were neither good conductors of electricity nor good insulators, but were somewhere in between. Further investigations of how semiconductors work, using the quantum theory of matter, laid the basis for the invention of the transistor. The transistor made possible the development of modern solid state electronics, which is the basis for modern computers.

We must beware of pseudoscience, of discoveries claimed by people who think they are greater than Leonardo da Vinci. There is much talk about telepathy, parapsychology, levitation and psychokinesis. Of course scientists today do not know all the answers. There may be new forces which we cannot imagine. There may be very different means of communication between people that we do not yet understand. But how do we look for them and find them? One thing we do know. Many great geniuses have been looking for these things over thousands of years and they have not found them. We are not greater nor cleverer nor luckier than Einstein and Leonardo. If these exotic phenomena are really there like Jupiter's moons, just waiting to be discovered, we will need some new tools to find them. The tools that Leonardo had were not enough.

Whenever anyone claims that he has made a great discovery that all these great people simply missed or didn't think of, I don't believe him. Much of it is nonsense. Much of it is just plain fraud. None of us is greater than Leonardo and all the others. If anybody can bend a spoon at a distance without cheating, and using only tools that were available to Leonardo, then Leonardo would have done it and written a treatise on how it was done. We must wait until the time is ripe for new discoveries, when all the tools and accumulated knowledge needed are available. We must realize that we can only add a very modest contribution to the heritage of accumulated scientific knowledge which was left to us by our predecessors.

Another misconception of many laymen is that there exists a "scientific establishment" which resists new ideas and that people with new ideas outside the establishment are stifled. The main problem in any good research institution is not a lack of new ideas but rather that there are too many new ideas and nearly all of them lead nowhere. A good scientist must know how to discard unfruitful ideas quickly. He cannot afford to spend all his time proving that bad ideas lead nowhere. He must concentrate on those directions which seem to show promise. Any new ideas must therefore be tested critically and proved wrong or unpromising as soon as possible. The same is true of experiments. Good experimental research is difficult in the areas where much is still unknown. It is very easy to make mistakes and get incorrect results. One can waste a career trying to understand experiments which later turn out to have been wrong all the time.

Suppose for example an experimenter in the laboratory finds a new effect which might mean a dramatic breakthrough. But such an effect could also be caused by tiny amounts of dirt or impurities in the experimental apparatus. The scientist has to decide whether the effect is significant or whether it is due to dirty apparatus. He can spend the next ten years improving his apparatus and making it cleaner and cleaner. But if the effect finally goes away, he will have wasted ten years of his career proving nothing except that dirty apparatus produces results which look exciting if they are not properly understood. But if he dismisses the effect as due to dirty apparatus without a thorough investigation when he was really on the threshold of a great discovery, he has missed the opportunity to make a dramatic contribution to human knowledge.

How is a scientist to know whether an effect is real or due to dirty apparatus? There are no rules and regulations given by the scientific establishment. The truly great experimentalist has a feeling for what is significant and what is dirt. It is here that experimental research is an art rather than a well-defined scientific procedure. The scientific method cannot tell the scientist whether it is worthwhile to follow a particular line of research or whether to drop it and look elsewhere. The precision of the scientific method can only help in drawing conclusions about experiments and for indicating what further experiments are necessary to decide whether an effect is real or whether it is dirt. But whether it is worthwhile to spend a week, a month, a year, five years or ten years in tracking down this effect cannot be decided on the basis of any well defined rules.

One example of a scientist who tracked down a great discovery was Rudolf Mössbauer, who as a graduate student discovered a tiny effect which was not anticipated. He had a radioactive source, a counter for detecting the radiation and an absorber placed between the source and the counter. He was studying the changes in the counting rate of radiation at the counter when he changed the temperature of the source and the absorber. This was supposed to give him information about the structure of the radioactive nuclei emitting the radiation. When he cooled the source to liquid air temperature the counting rate went down as predicted by the theory. When he cooled the absorber the counting rate also went down. But when he cooled both the source and the absorber down to liquid air temperature at the same time, the counting rate did not go down even more as expected but went up very markedly instead. The whole effect was small, less than one per cent. All kinds of "dirt effects" could produce such a small change in counting rate. Drastic cooling causes thermal contraction and could change the sizes of different parts of the apparatus. Many other things might go wrong and produce such an effect. Yet Mössbauer was able to follow up this effect, find other ways of testing it to demonstrate that it was real, and to begin a new promising direction of research. This work opened a new field with applications in many areas remote from the original problem of nuclear structure where the effect was discovered.

Mössbauer received the Nobel Prize for this work which began with a perplexing experiment and a gifted experimentalist who had the ability to distinguish between a real effect and dirt.

From Relativistic Quantum Theory to the Human Brain

In October, 1984 we heard of the death of Prof. P. A. M. Dirac, one of the giants of twentieth century physics. Shortly afterwards a conference on "The Impact of Science on our Lives" at the Weizmann Institute heard a lecture on the latest research into the human brain, investigating the biochemical effects of anti-depression drugs on the operation of the brain. A key ingredient in this brain research was the use of positron emission tomography (PET), a technique for looking deep into the brain with the use of radioactive atoms which emit particles called positrons. The positron, discovered a half century ago, is a very unusual particle called the antiparticle of the electron. This leads us back to Dirac, who predicted the existence of the positron a number of years before its discovery. I shall now attempt to tell the fascinating story of the positron from Dirac to brain research.

In the 1920's Dirac played a very important role in the development of the new revolutionary quantum theory of the atom. The quantum theory explained many of the puzzling features of atomic physics, which could not be understood with the nineteenth century mechanics of Newton. But the other great revolution of the beginning of this century, Albert Einstein's theory of relativity, was not incorporated into the new quantum theory. They seemed to be very different.

The quantum theory dealt with the failure of Newtonian mechanics to describe the motion of very tiny objects the size of atoms, and provided a revolutionary new approach to atomic phenomena. Einstein dealt with the failure of Newtonian mechanics to describe the motion of objects at very high speed, like the motion of the Earth around the Sun and the passage of light rays near the Sun's surface. Einstein provided a revolutionary new approach to motion at high speeds. All attempts to put the two great revolutions together encountered formidable difficulties. Dirac was disturbed because the quantum theory, on which he had worked so hard, was not consistent with relativity. He set himself the task of finding a way to combine quantum theory and Einstein's relativity and using his new approach to describe one of the basic particles of nature, the electron.

Today we are all familiar with the picture of the atom, with the electron moving in an orbit around the nucleus like the planets around the Sun. Electrons also move freely from one atom to another in materials called electrical conductors. This motion produces the electric currents that bring us light, telephone messages, heat from electric heaters, and power from electric motors. The electrons moving in individual atoms and moving from one atom to another in radio antennas and microwave cookers radiate the electromagnetic waves which appear to us in many different phenomena like light, radio and television signals, heat and laser beams. The new quantum theory of the 1920's explained all these properties of the electron, but completely ignored Einstein's relativity. This seemed very reasonable at the time, because all these phenomena involved only slowly moving electrons and the new effects of relativity only appeared when particles moved at very high speed, near the speed of light. For slowly moving particles, Einstein's equations of motion were very nearly the same as Newton's equations of motion.

Dirac felt that it must be possible to combine relativity with quantum theory, to obtain a new theory which would also describe the motion of very rapidly moving electrons. He developed an equation which described all known properties of the electron and incorporated the principles of both the quantum theory and Einstein's relativity. It described all the electron orbits observed experimentally in the hydrogen atom very precisely and in excellent agreement with the results of experiments, including very fine effects not previously described with the old nonrelativistic theory.

But Dirac's relativistic quantum theory of the electron had a very peculiar side effect. It also described a new family of very crazy electron orbits which seemed to make no sense at all. Furthermore, Dirac's theory implied that an electron moving in one of its regular orbits in an atom could suddenly jump into one of these new crazy orbits and at the same time release an enormous amount of energy. Nobody understood the meaning of these crazy orbits appearing in a beautiful new theory which was otherwise so sensible.

Several years later Dirac himself proposed the answer to the puzzle of the crazy orbits. He postulated the existence of a new particle, the positron, which was the antiparticle of the electron. It had all the same properties as the electron, except that it had the opposite sign of electric charge. The charge of the electron is negative; the charge of the positron is positive. Dirac's equation described the orbits of both the electron and of its antiparticle the positron. The crazy orbits that had perplexed everyone suddenly became reasonable when they were interpreted as orbits of a new and different particle which had the opposite sign of electric charge from the electron.

The enormous energy released when an electron jumped from a regular orbit to a crazy orbit also had a simple and revolutionary interpretation in this new picture. Einstein's theory of relativity said that matter and energy were related; that matter could be converted into energy and energy into matter. Dirac's equation described precisely this conversion. An electron and a positron could annihilate one another and their mass would be converted into energy. The energy of an X-ray could be converted into matter by creating an electron and a positron together. The jumping of an electron into a crazy orbit was simply the annihilation of an electron against a positron in the crazy orbit.

Dirac's revolutionary breakthrough introduced the new concept of antimatter. For every particle, there was an antiparticle which had the opposite electric charge, and a particle and an antiparticle could annihilate one another and turn their mass into energy. This was an inevitable consequence of combining Einstein's relativity with the new quantum theory. Soon afterwards Dirac's prediction was confirmed by experiment. The positron was discovered in the cosmic rays entering the earth's atmosphere from outer space. Then many radioactive nuclei were found which emitted positrons. Many years later the antiparticle of the proton, the antiproton, was discovered. Today the existence of antiparticles for all particles is accepted and agrees with all experimental results.

Hundreds of radioactive nuclei which emit positrons are now known, and some of these have been found to be useful in many applications, like brain research. In the work described in the meeting at the Weizmann Institute, an atom with a radioactive nucleus was used which behaved chemically like an atom naturally attracted to a particular chemical found in the brain. When such radioactive atoms were injected into a specimen of brain tissue, they were naturally attracted to this chemical. The brain specimen was then placed in an instrument called a positron emission tomograph which detects the energy released when a radioactive atom emits a positron, computes exactly where this occurred in the specimen and produces a picture called a PET-scan which shows the precise point where any radioactive atom and the chemical that attracted it had been. The result is a precise map of the brain showing where the particular chemical related to antidepression drugs appeared. By studying the brains of animals which had undergone different treatments, it was possible to pinpoint the effects of these drugs.

The energy which produced these pictures came from the annihilation of the positrons against the normal electrons which were present in the material. Positrons are not usually found in the matter we see every day, which consists of atoms built from nuclei and electrons. A positron emitted into any normal material quickly finds an electron, annihilates it, and emits the energy as two gamma rays (similar to X-rays), each with a very large energy of about one half million electron volts. The gamma rays make it easy to detect positrons, since no other known phenomenon produces two gamma rays emitted at exactly the same time in opposite directions with exactly this high energy.

Sophisticated instruments like the positron emission tomograph use these gamma rays to detect the annihilation of a positron against an electron in a specimen of matter and pinpoint the exact spot where the annihilation occurred. These instruments contain gamma ray detectors which convert the gamma ray energy into a short pulse of electrical energy, measure the gamma ray energy and record the exact time and place at which the gamma ray was detected. This information is fed into electronic computer circuits which search for cases where simultaneous pulses of electrical energy are observed in two gamma ray detectors on opposite sides of the specimen and check that the gamma rays have exactly the right energy. This shows that two gamma rays were emitted exactly at the same time in opposite directions from the specimen. The observation of *two* gamma rays emitted in exactly opposite directions and observed in two different detectors in different places provides the information that enables the computer to determine the exact location of the radioactive atoms that emitted the positrons.

This kind of information is obtainable only from radioactive atoms that emit positrons. Other types of radioactive atoms also emit radiation that can produce energy pulses in detectors, but a single pulse in one detector only tells us that radiation hit that detector. It provides no information about the direction from which the radiation came. The simultaneous pulses in two detectors produced by positron annihilation tell us also the direction. The source of the radiation must have been somewhere between these two detectors and along a straight line connecting them.

The energy of these gamma rays from positron–electron annihilation, one half million electron volts, is enormous compared with the energy of only a few electron volts released when an electron jumps from one normal orbit to another in the hydrogen atom. But it is just the mysterious energy which Dirac's original equation showed was released when an electron jumped from a normal orbit to a crazy orbit. The peculiar side effect of Dirac's theory, when properly interpreted, has now not only been confirmed by experiment, but has found uses in industry, medicine and research which no one could have anticipated at the time when Dirac predicted the positron.

Another side effect of Dirac's equation was its prediction of the spin and magnetic properties of the electron. The electron, like many other elementary particles spins like a top and behaves like a tiny magnet. Dirac's theory predicted this behaviour and gave the exact and correct values of the spin and of the strength of the magnet.

In 1974 Dirac was asked to review the history of these developments at an international conference on Spin Physics at the Argonne National Laboratory near Chicago. Someone in the audience asked him if he had been disturbed when he first disovered that these crazy orbits of the electron had appeared in his equation. His answer was very interesting and instructive, shedding light not only on the workings of the mind of a great physicist, but on a general approach to frontier research into the nature of matter and energy.

Dirac said that he had set himself the goal of resolving the difficulty of reconciling the two new great revolutions of quantum theory and relativity. He had worked hard to achieve this description and he had succeeded. But, said Dirac, one can never hope to resolve all the difficulties at once. It is natural that in the process of resolving some difficulties, other new difficulties arise. This simply sets the stage for the next research project, the resolution of the new difficulties. This is the way that progress is continually made in our understanding of nature.

The prediction of the electron's spin and the strength of its magnet came as a complete surprise to Dirac. He had only wanted to combine relativity and quantum theory and had thought that it would be easiest for a particle that did not spin and had no magnet. He felt that spin and magnetism could be added later on; he only wanted to solve one problem at a time. But they came as a free bonus in his theory.

Dirac's relativistic equation of the electron solved the difficulty of bringing relativity and quantum theory together. It also solved the problem of the spin and magnetism of the electron, which Dirac had not expected. Its crazy orbits introduced a new difficulty, which was later solved by Dirac, leading us into a new world of matter and antimatter, transformation of energy into matter and matter into energy, and eventually to radioactive nuclei useful in brain research. And these new concepts led to many new difficulties, whose resolution have led to more difficulties.

The chain of events started by Dirac's revolutionary discovery still continues, as physicists discover more and more new particles and attempt to understand the nature of the basic building blocks of the world we live in. Alongside of these developments the biologists and biochemists have been using the side effects of Dirac's discovery to create new tools and new techniques for investigating living organisms and understanding the human brain.

Appendix

The Impact of Dirac's Positrons on My Own Career

My own scientific career began with Dirac's positrons. In 1946, when I started my graduate study at Princeton, the positron had already been discovered, and the transformation of matter into energy when a positron collided with an electron had been well established. But there had not yet been any experimental confirmation that rapidly moving positrons behaved in accordance with Dirac's equation. Everybody believed that the equation was correct; it had been tested for electrons and the positron was the antiparticle of the electron. Still physicists are not satisfied until every possible loophole has been checked and theoretical predictions are directly confirmed by experiment.

One way to check Dirac's theory is to study what happens when a rapidly moving positron passes close to a the nucleus of a heavy atom like platinum. The electric force between the nucleus and the positron deflects the motion of the positron from its original direction, and the amount of deflection can be measured in a laboratory experiment. Dirac's equation predicted that the amount of deflection would be different from the amount predicted by the old theory. This difference results both from the new effects of Einstein's relativity and from magnetic forces acting on the positron magnet predicted by Dirac. The British theorist N. F. Mott had investigated this process for both electrons and positrons passing close to a nucleus, using Dirac's equation, and had published exact quantitative predictions for results of experiments. This process, now called "Mott Scattering", could be observed in the laboratory by shooting a beam of positrons at a piece of platinum and looking for positrons coming out of the platinum in different directions.

Such deflection experiments were first performed by the famous British experimenter Ernest Rutherford. He investigated the internal structure of the atom by measuring the deflection of particles called "alpha particles" when they passed through atoms. Rutherford calculated how particles are deflected by electric forces. This process has since been called "Rutherford Scattering" and the formula that gives the exact amount of the deflection is called the Rutherford formula.

Rutherford's experiments found very large deflections and proved that the interior of the atom was very different from a homogeneous mass like a lump of jelly as many physicists believed at that time. Rutherford's formula showed that such large deflections could not be produced when an alpha particle passed through that kind of atom. But if nearly all of the mass of the atom is concentrated in a tiny nucleus with electrons moving in large orbits around it, the alpha particles can easily pass through the empty space between the electrons and the nucleus, come very close to the nucleus and be strongly deflected by the strong electric forces. This is how the modern picture of the atom was discovered.

Rutherford's experiments used slowly moving particles, where the effects of relativity were negligible. The old "nonrelativistic theory" used by Rutherford predicted that electrons and positrons would both be deflected by the same amount when they came near a nucleus. Mott's prediction from Dirac's theory predicted that the effects of relativity and of the magnetic forces were very different for electrons and positrons. The "Mott scattering formula" showed that when electrons and positrons were moving very rapidly at speeds very close to the speed of light and very close to a platinum nucleus the deflection of an electron could be three times stronger than the deflection of a positron under the same conditions. This had not yet been checked when I started my graduate study in 1946. In my Ph. D. thesis research project I compared the deflections of positrons and electrons emitted from different radioactive nuclei when they passed through thin foils of platinum. The results agreed with Dirac's theory, and nobody was surprised.

When I was starting my thesis work, my thesis adviser Prof. M. G. White suggested that I look into a different test of Dirac's theory. An Indian theorist H. Bhabha had studied the process of collisions between positrons and electrons using Dirac's equation. He found that when a positron came close to an electron, there was another process that could occur in addition to the deflection of the positron by electric and magnetic forces. The electron and positron could annihilate one another and turn their mass into the energy of gamma radiation, but then the gamma rays could turn back into matter again and create an electron and a positron moving in a different direction.

Prof. White suggested that I try to measure this process, which is now called "Bhabha scattering". The effect could be observed in the laboratory by shooting a beam of positrons at a target material containing electrons, and looking for positrons coming out of the target in a different direction. Some of the positrons detected would have simply been deflected by the electromagnetic forces, but there would be an additional contribution from the annihilation of the electron–positron pair into gamma ray energy and the creation of a new electron–positron pair.

I found that the experiment was not feasible at that time. Even if I used the strongest possible source of positrons available at the time, and the best possible detectors, I would not get enough positrons into my detector to give a significant effect. However, I then noted that the predictions from Dirac theory for Mott scattering had not yet been checked, and found that this experiment was feasible and was a suitable subject for a thesis.

Much has changed since those days. The best detectors available for positrons and electrons were Geiger counters. A few doors down from my laboratory in the basement of the Princeton physics building, a young professor named Robert Hofstadter was investigating the possibility of using sodium iodide crystals as detectors for radiation. But these still required extensive development before they could be used in any experiment. The best source of positrons I could obtain came from a radioactive nucleus which was made by shooting a beam of helium nuclei into a piece of copper in a machine called a cyclotron. In collisions between the helium and copper nuclei radioactive nuclei of the element gallium were formed which decayed by emitting high speed positrons. The positron sources for my experiment were prepared in a cyclotron in Washington, D.C. by bombarding a copper target with helium nuclei for a full day. In the evening the radioactive target was flown to Princeton by a private plane. A radiochemist then separated out the radioactive gallium from the lump of copper and deposited it into my apparatus. By midnight I started my experiment, and worked night and day for several days, until my radioactive source had lost its activity. The particular nucleus I used had a nine hour "half life", which means that it loses half its strength every nine hours. After 36 hours it was too weak to use any more.

Today Hofstadter's sodium iodide crystals have been developed and are widely available as radiation detectors. There are machines for producing intense beams of positrons and electrons, and collisions between very high speed electrons and positrons are studied with the aim of finding new particles. One of the detectors used with these colliding "electron–positron" accelerators is called the "crystal ball" and consists of tons of sodium iodide.

There has been tremendous progress in the development of particle physics during Dirac's lifetime. His equation and Bhabha scattering are now well established; nobody worries about testing them any more. On the contrary, the Bhabha scattering which was considered so interesting but undetectable in 1948, is easily detected and no longer of interest in 1984. Instead it is always present as a "background" in the experiments looking for other new effects of greater interest in electron–positron collisions. Sometimes Bhabha scattering is measured as a convenient way to check whether the apparatus is working properly.

Creative Questioning

Someone once asked a Jew: "Why do you Jews always answer a question by asking another question?" The Jew answered "Why not?".

This questioning which is so much a part of the Jewish tradition is also the key to scientific progress. The creative scientist is not looking for the final scientifically proven answer to important questions. He is investigating interesting questions with the hope that they will lead to even more interesting questions. The universe and the knowledge to be discovered is boundless. What we don't know is much more exciting than what we already know.

Newton discovered his laws of motion and gravitation by asking questions about the motions of bodies and the forces that make them move. He learned that the laws of force that describe the falling of an apple from a tree also describe the motion of the moon around the earth and of the earth around the sun. The questions that Newton asked and answered to the best of his ability led us to new knowledge and new questions. The students who studied Newton's laws in school were led to a greater understanding of the universe and used Newton's laws for the developments and inventions that made modern technology possible.

The debate on science teaching and evolution misses the point completely when it gets involved with the scientific proof of the theory of evolution. Of course there is no scientific proof of evolution. There is also no scientific proof of the theories of atomic physics, electricity and magnetism, or Newton's laws of motion or gravitation.

In fact, Einstein showed that Newton's laws were not precisely correct. But fortunately our schools taught Newton's laws even though they were not scientifically proven and in fact turned out not to be precisely correct. Today we have Einstein's new laws of motion and gravitation which refined Newton's laws and give a more precise description of motion and gravity. But we still teach Newton's laws in our schools because they are much simpler and adequate for all practical purposes. Most students will never need to learn Einstein's theories of special and general relativity, while they will find the basic principles of Newton's laws useful in everything from driving a car to astronautics.

The modern theory of what goes on inside the atom is still questioned on philosophical grounds by many people. Einstein never accepted it. But nobody has found a better theory, and it is the only one that tells us how to use the energy of the atom, how to make transistors, how to build modern computers and how a laser works. If we waited until this quantum theory of atomic physics was "scientifically proven" before teaching it to students, we would not have transistors, computers and lasers today.

A true scientist is always open to new questions and new knowledge. If an alternative theory to evolution is presented which is manifestly better, it will be accepted, but only after exhaustive criticism and questioning. A great new theory not only answers questions and poses new ones; it also shows new relations between phenomena which were previously believed to be completely different. Newton's laws not only explained the falling apple; it showed the relation between falling objects on Earth and the motion of the Earth itself. Maxwell's theory of electromagnetism showed that electricity, magnetism, light and radio waves were all different aspects of the same basic fields of force. Einstein's theory of general relativity showed that inertia and gravity were intimately related.

Darwin's theory of evolution was accepted by the scientific community only after a long period of intensive criticism. It has proved its value by showing the way to new questions and by relating different phenomena which were otherwise completely unrelated. The same theory which explains all kinds of fossils and bones of prehistoric animals found in the Earth also explains the evolution of bacteria. The scientists developing new drugs and antibiotics and the doctors who use them to treat diseases also use the theory of evolution to explain how new strains of bacteria can evolve that are resistant to drugs and to help them find new ways to keep the drugs effective. An alternative theory must be able to do everything that the accepted theory does, and more. So far no such theory has been presented. Criticizing the presently accepted theory with hairsplitting arguments is pointless. All accepted scientific theories have their flaws; none of them is perfect. If there is something that is better than evolution, the scientific community must be convinced that it is better.

The theory of evolution is an integral part of modern biology which has led to so many of the advances in modern medicine. The biology students who learned this modern approach in school went on to develop antibiotics, vaccines that eliminated diseases like polio which crippled so many of our children, open heart surgery and hopefully to find the cure for cancer. Allowing religion or ideology to interfere with teaching our students the knowledge that we have has inevitably led to disaster. Soviet biology took a long time to recover from Stalin's support of the charlatan Lysenko, who attempted to impose ideology on science. Christian Scientists in America who allow religious prejudices to interfere with their use of modern medicine are denied its benefits.

Creationism cannot cure cancer. Creationism cannot feed the hungry. Creationism cannot solve the energy crisis. But the study of evolution may lead to the solutions of these important problems facing mankind. Scientists may use their knowledge of evolution to develop new strains of bacteria, viruses or other living organisms to work for the benefit of the world.

Somewhere in our schools there may be a young student who will find the cure for cancer or develop new strains of plants that revolutionize agriculture and feed the hungry people of the world. Our schools must give him the tools he needs for his future work by teaching him the best of our secular scientific knowledge. There is no shortage of religious education in Israel today, and any student who is interested can learn everything written in the Bible about the creation. We should not mix religion with science.

5

The Building Blocks of Matter — What is a Quark?

To begin this chapter I should explain what a quark is, and why we are interested in them. The place to find out what a word means is the dictionary. But any dictionary will not do. You have to know the language. The word quark can be found in any German dictionary, and this is a typical example of what one finds there.

Quark, m. curd, curds; slime, slush, filth; (fig.) trifle, rubbish, trash. . . . . .

So why should anyone be interested in this kind of junk? Perhaps this chapter will explain it.

Modern atomic physics began with the discovery in the early part of this century that matter which appears so solid and massive to the layman really consists mainly of empty space with tiny particles rushing around at very high speeds. Today we all know that matter is composed of atoms which look like the familiar picture of a miniature solar system with electrons orbiting around a nucleus. We also know that there are tremendous energies locked in the nucleus. But although scientists and engineers have learned how to release and to use some of this nuclear energy, they still do not understand the structure of the nucleus and the many peculiar particles that are found inside. These particles are much too tiny to be observed with the most powerful microscopes and can only be studied with machines called "High Energy Accelerators". These machines produce beams of tiny particles moving at extremely high speeds. By studying what happens when these high speed particles collide with other particles physicists endeavor to unravel the secrets of their structure.

The use of collisions between high speed particles to study the nature of matter was developed by Rutherford and enabled him to discover that the atom consisted of a tiny nucleus surrounded mainly by empty space, rather than being continuous and homogeneous as matter appears to the naked eye. Today physicists use the same technique to probe the interior of the nucleus, using beams of much higher energy particles. Sometimes such high energy collisions create new particles which were not known before. Physicists thought at first that these particles must be the fundamental building blocks out of which all matter is made. But by 1960 so many different new particles had already been discovered that physicists began to think that perhaps these particles were made from even tinier particles arranged in different ways. It was found that a large group of particles which had been discovered had properties which suggested that they were all made by putting together three basic building blocks, which were given the name "quarks". Extensive searches for quarks were undertaken since then by physicists who had hoped to discover the ultimate building block from which all matter was made. But so far no isolated quarks have been found. Instead more and more other particles were discovered which all look as if they were in some way made out of quarks.

One day a few years ago I walked into my dentist's office and looked through an old German magazine while waiting. I was surprised to find a quark staring back at me in a full-page advertisement showing a big picture of a potato with an enormous topping of a white cheese, and the legend "Hmm — Kartoffel mit Quark — Aus Deutschen Landen Frisch auf dem Tisch" (Hmm — Potato with Quark — from German lands fresh on the table). So this is the ultimate building block of nature! When I was a boy, people used to say that the moon was made of green cheese. Now man has gone to the moon and we know better. Today some physicists say that all matter is made of these quarks. Perhaps in another few decades we will know better about this, too.

By this time you must be thoroughly confused about how the cheese gets into these particles. The name quark was actually taken from James Joyce's book *Finnegan's Wake* by Murray Gell-Mann, who was one of the first to suggest that these building blocks might exist. Gell-Mann did not realize at the time that it was a German word; he thought it had been coined by Joyce. By the time the true origin of the word quark became widely known, it was too late to change it to something more dignified, and many of the physicists who did not believe in quarks thought the name was appropriate. So the name has remained. But so far nobody has found any.

It is now accepted that quarks exist. But although many experiments have told us that nucleons and nuclei were made of quarks, no experiments have been able to find any isolated quarks. There were many unsuccessful experimental searches for quarks, including in places like sea water and meteorites. Experiments with high energy collisions between protons had been expected to break up protons into their constituent quarks. Instead these collisions produced a complicated mess of many particles which nobody could unravel and no isolated quarks.

We now believe that the forces binding quarks into a proton are so strong that enormous energy would be needed to knock a quark out of a proton. These building blocks are combined in groups of two or three by such strong forces that it is impossible to separate out single quarks. Physicists have tried to break these groups of building blocks into quarks by accelerating them to produce very high speed collisions. But they are defeated here by Einstein's principle relating energy and matter. At very high speeds they have so much energy that the energy is turned into making more building blocks.

We can never create a single quark just as we can never create a single electron. An electron has a negative electric charge and can only be created together with another particle like a positron that has a positive electric charge. Similarly a quark has a definite color charge and can only be created with an antiquark which has an opposite color charge. But the electric force between electron and a positron decreases with distance just like the gravitational force between a rocket and the Earth. If a rocket is hit hard enough it can escape from the Earth. In the same way the electron can get away from the positron and be observed as a single electron.

But the very strong forces between a quark and an antiquark can never allow a single quark to escape. When a collision between two protons has enough energy to knock a quark out of the proton, the energy is converted into new quarks. A quark–antiquark pair can be created in the space between the struck quark and the two other quarks from the proton. The antiquark is then captured by the struck quark to make a meson and the new quark joins the two old quarks to make a new proton. The isolated quark is not observed, only a cloud of new particles.

The Structure of Matter

What is the structure of matter, and why are such accelerators necessary to study it, rather than ordinary microscopes? To get some feeling for the answer to these questions, let us first consider how scientists (or laymen, for that matter) study the structure of more familiar objects. When we look at a brick wall from a distance, it is hard to discern that it is made up of individual bricks and cement. From far it looks as if it were made of one homogeneous material. Only by coming closer to it, or looking at it through binoculars or a small telescope, can we see its fine details and establish that it is constructed of building blocks arranged in a very special and regular pattern to provide the stability and strength that a wall needs, and that there is cement between the bricks that makes them stick together. If we look at the bricks even closer, we see that each individual brick has a structure and is made of even smaller constituents. In the same way, a bar of iron that appears to be a homogeneous structure is really made up of atoms arranged in a specific pattern to give iron its special hardness and smoothness.

The study of the structure of matter is like the study of a brick wall at many different levels. The brick wall appears to be homogeneous, but it is built out of building blocks arranged in a regular pattern. The bricks are held together in a wall by cement which hold them together. And the bricks themselves are kept in shape by their own internal strength and structure. We can ask these three same questions about solids, molecules, atoms, nuclei and even the tiniest known particles.

Are these objects made of smaller building blocks? If so what are these building blocks?

How are the building blocks arranged and what are the forces like cement or glue that hold them together?

What is the structure of the building blocks? What are the forces that keep them in their shape like a brick instead of being squeezed together like a lump of jelly?

Scientific research into the structure of matter follows an approach based on these answers to the three questions:

1. The simplest way to see if an object is made of smaller building blocks and to identify the blocks is to break it up. If we hit a brick wall hard enough with a sledgehammer in the right place, we can knock a brick out of it and examine the brick. In the same way scientists have broken up solids, molecules, atoms and nuclei and found that they were made of smaller building blocks. But this brute force method destroys the structure of the original object. If we want to study how the bricks are fit together in a wall and what holds them together, we need finer tools than a sledgehammer.

2. We can use light to study the structure of the brick wall without breaking it up and see how the bricks are arranged and the forces that hold them together. We examine a brick wall with light that comes from a source like the sun, hits the brick wall and is reflected into our eyes. The picture we see of the brick wall depends not only on its structure but also on the kind of light available to us. The most ordinary wall looks quite different when it is lit by sunlight, by

artificial light, by red or green light. Infrared or ultraviolet light give us still different pictures, and a picture of a wall taken with X-rays bears little resemblance to that taken with visible light.

And light itself has a structure. It is not one single continuous beam, but rather a hail of tiny bullets, called photons, which hit the wall, bounce back and convey the message to us that the wall is there. As long as the objects under investigation are much bigger than these photon bullets, we needn't worry much about the structure of the light, just as we can use iron to make the most complicated machinery without having to know or really care that in reality it is made up of separate atoms. But objects the same size as the light bullets or even smaller will give blurred pictures. For a clear and informative picture, tinier bullets are needed to study very small objects. But we must take care that our barrage of light bullets is sufficiently gentle and does not destroy our specimen like a sledgehammer.

3. The study of the bricks themselves leads us to the realization that the stability of a brick wall comes from a balance between two opposing kind of forces. There are the attractive forces of the cement which hold the bricks together. But there must also be other forces which prevent the bricks from collapsing under the weight of the wall, as would happen if the bricks were lumps of jelly. Trying to understand these forces leads us back to the first question, now at a tinier level. Instead of examining the structure of the wall and finding that its building blocks are bricks, we examine the structure of the individual bricks and ask what its building blocks are.

These three questions thus keep repeating themselves over and over again at tinier and tinier levels. Will this ever stop? Will we reach a final tiniest level or does it go on forever? Today nobody knows the answer.

An accelerator is used both as a sledgehammer to break up the nucleus and as a beam of particles like a beam of light which can bounce back from the nucleus without destroying it and give information about its structure. The accelerator produces a beam of very high speed particles which is focused on the material being studied. When such a high speed particle hits a nucleus in a head-on collision, it breaks up the nucleus like a sledgehammer knocks bricks out of a wall. The physicists then study the pieces that come out of the nucleus to determine what its building blocks are. When such high speed particles come near enough to the nucleus to make a glancing collision without breaking it up, the physicists can study the particles bouncing back and get a "picture" of the nucleus like the picture of a brick wall that we get from illuminating it with bullets of light and observing the ones that bounce back into our eyes or into a camera.

So far we have overlooked one important aspect of the submicroscopic world. Unlike bricks in a wall which don't move very much, the atoms in the subatomic world move constantly at higher and higher speeds. We have all seen the atom depicted as a miniature solar system, with electrons moving around a nucleus in orbits that resemble those of the planets around the sun. The nucleus, which appears as a dot in the center of this familiar picture, actually contains neutrons and protons that move even faster inside the nucleus than the electrons on the outside.

The discovery in the early part of this century that tiny atoms move at incredibly high speeds and high energies revolutionized scientific concepts about the structure of matter and the sources of energy. Today everyone knows that tremendous energies are locked in the tiny atom and that even tinier objects have even larger energies.

The high speed also helps answer the question of what helps the bricks keep their shape instead of being squeezed together like lumps of jelly. The bricks are made of atoms and molecules which are held together by strong attractive forces that keep the brick from falling apart. But what keeps the atoms far enough apart so that they make a brick instead of something much smaller? We can ask the same question about the Earth and the Sun. The Sun attracts the Earth with a very strong gravitational force. What keeps the Earth from falling into the Sun?

We can see the answer to this question by noting what happens when we throw an object up into the air. It goes up for a while and then comes down. The gravitational force of the Earth slows it down until it stops moving upward and then pulls it down to the Earth. The faster it goes, the farther away it will get before it comes down. But if we shoot a rocket with enough speed, it can overcome the Earth's gravity and get away from the Earth. Once it is moving fast enough, we can give it a kick sideways with another rocket blast and make it into a satellite that moves around the Earth without falling into it. Its motion keeps it from falling into the Earth. The faster it moves, the farther away it can get from the Earth. The size of its orbit depends upon its speed.

It is this same kind of motion that keeps atoms apart in solid matter like bricks and gives the bricks their shape and strength. The attractive forces which want to pull the subatomic particles together are like the force between the Earth and the Sun or the satellite and the Earth. But the rapid motion of all these particles prevents them from falling into one another and determines the size of their orbits.

How Accelerators are Used as Microscopes

Let me now try to explain how scientists use accelerators to study the nature of particles which are so small that they cannot be seen even through the most powerful microscopes. The familiar picture of the atom as a miniature solar system with electrons moving around a nucleus like planets around the sun is based on experiments carried out in England by Lord Rutherford in the early part of this century. Before Rutherford's experiments, physicists thought that the atom had a very different structure, something more like a homogeneous lump of jelly. How is it possible to decide whether something so small is a miniature solar system or a lump of jelly? To get an idea of how this is done, let us imagine that our solar system is being studied by a race of giants far away in outer space. They have indications from their observations that there is some object in this region of space, but they cannot tell whether it is a huge lump of jelly, a huge cloud of dust, or something else. They cannot come close and simply look.

Their only way to explore this region of space is to shoot rockets at the solar system and follow their path in space.

The giants understand Newton's laws of motion and gravitation, and they know that the rocket will follow a straight line course in space as long as it is far away from any matter. But when the rocket passes near the solar system, a gravitational force deflects it away from its straight line course. The giants have precise instruments for studying the orbit of the rocket. From the data on the orbit they can calculate how big the forces are that act on the rocket and where they come from. Since they cannot see the solar system, they cannot aim their rockets very precisely at its center. They shoot off millions of rockets and follow all the orbits. Some of them will miss the solar system by large distances and not feel any gravitational pull at all. Others will come very close, feel a strong force and be sharply deflected. By studying how many are deflected and how much they are deflected, they get a picture of the solar system.

Nearly all the matter in the solar system is concentrated in the sun, and the gravitational force felt by the rockets would come almost entirely from the sun. Newton's laws tell us that the motion of a rocket will follow the simple orbits that are easily calculated as long as the rocket does not enter the sun itself. But any rocket which enters the sun will immediately be destroyed by the sun's heat and disappear. So the giants trying to study the solar system by shooting rockets at it would find that there is a very large mass of matter concentrated at the center, that rockets are deflected by it, and that every so often a rocket gets lost, indicating that it hit something. They could measure the size of the sun by studying the amount of deflection of different rockets. They know that rockets which come very close to the sun are deflected more than those which are far away. The ones which are deflected the most are those which just miss the surface of the sun and then get away. Any rocket which comes closer would be lost. They would look for those rockets which are deflected by the largest possible angle, and keep looking for rockets which might be deflected even more. After they have shot very many rockets and were sure that there were none that were deflected more, they would conclude that the rockets that had the maximum deflection had just missed the surface of the sun. They could then calculate the orbit of the rocket and calculate the size of the sun.

Suppose that the solar system had most of its matter spread out in a large cloud of dust, instead of being concentrated all in the sun. Then the rocket orbits would follow Newton's simple theories only as long as the rockets didn't get into the cloud itself. Once a rocket got into the cloud, all kinds of things might happen, depending upon the speed and material of the rocket and the material and structure of the cloud. The rocket might be slowed down by friction, but still get out. It might get burned up and be lost. If the dust clouds contained big rocks about the same size as the rocket, the rocket might bounce off the rock and go off in a completely different direction. It might be split in two by a collision with a rock. The giants might design special rockets to look for such rocks, one which would split in two on impact, but still send back messages about its orbit. They might design rockets of different materials and find that rockets of one material burned up and got lost, while rockets of another material got through. And they would still be able to measure the size of the dust cloud by looking for those rockets which just missed the edge of it. These would come out without any signs of collision and would have the greatest deflections.

When Rutherford did his first experiments shooting alpha particles at atoms, the atoms might see Rutherford as a giant shooting rockets at them. The alpha particles are high speed particles emitted from radioactive nuclei. The forces between alpha particles and atoms are not gravitational forces, but electrical forces. However, Rutherford knew the laws of motion of particles under electric forces and was able to calculate the orbits of the alpha particles just like the giants in our fictitious example could calculate rocket orbits.

Physicists believed that matter was made of atoms packed together like bricks in a wall, They knew the size of the atom, which they thought was a homogeneous piece of matter like a brick or a lump of jelly. Rutherford was able to calculate how much an alpha particle would be deflected if it came just close enough to the atom to miss it. To his great surprise, he found that many of his alpha particles were much more strongly deflected. They must have come much closer to the center of the atom without actually hitting the atom. In other words, the atom must consist mainly of empty space with all its electrical charge and its matter concentrated at its center in a very tiny region. The nucleus of the atom, in which the electric charge was concentrated, had to have a radius ten thousand times smaller than the size of the atom to account for the large deflections he observed.

Much of the research which attempts to investigate the structure of the tiniest constituents of matter today is simply an extension and a refinement of Rutherford's method. Instead of using beams of alpha particles from radioactive sources, we now have machines which produce more intense beams of particles having much higher energies, a million times higher than the energies of Rutherford's alpha particles. In this way we are able to study not only how big the nucleus is, but what is going on inside the nucleus. And like the giants who could study a dust cloud in space by using rockets made of different materials, physicists use different kinds of particle beams to study the interior of the nucleus. Electrons can pass through the nucleus, be deflected by electrical forces and still get out. When an electron is inside the nucleus, it has forces acting on it from all directions. Since we know the laws of electrical forces, we can use electrons to tell us the shape as well as the size of the nucleus, and also how the electrical charge is distributed inside the nucleus.

In addition to electrical forces, whose laws we know, there are mysterious very strong and very weak nuclear forces, which we still do not understand completely. These are the forces which hold the particles of the nucleus together. We can study the strong forces by shooting beams of particles like protons which experience these strong forces at a nucleus. Other particles can split in two while going through the nucleus and give more detailed information about what is going on inside. Then there are high energy collisions between particles in which the two coalesce and a completely new particle is produced. Many new particles have been discovered in this way. These new particles can then be shot at other nuclei to study new interactions. One of these particles is the neutrino, which has no electrical charge and does not exert any strong or electrical forces on the nucleus. The only forces it feels and exerts are the mysterious very weak forces. Physicists study the weak forces by shooting neutrinos at a nucleus.

More clues about the structure of the nucleus can come from shooting particles at it with different speeds. With the simple picture I have shown you, it would seem that the faster a particle goes by the nucleus, the less time it feels the forces from the nucleus, and the less it should be deflected. One would think that using higher and higher energy accelerators to shoot faster and faster particles at a nucleus would make it much easier for the particles to get by or through the nucleus with little happening to them. This had in fact been observed in experiments until recently. The faster the particles were, the easier it was for them to get by the nucleus. But then, at energies of around fifty billion electron volts, a peculiar thing happened. As the speed of the particles was increased even more, the chance of their being captured by the nucleus increased instead of decreasing further. This phenomenon has been very perplexing to physicists and is now one of the paradoxes which we are trying to understand. Why is it that a nucleus has a good chance to catch a particle going by when it is moving very slowly, less of a chance when it goes a little faster, but a better chance again when it is going even faster? New experiments at even higher energies and new ideas will be necessary to answer this question.

Gaps in the Mendeleev Periodic Table

In the search for smaller and smaller building blocks from which all matter is made, there has repeatedly been a stage where there are very many different kinds of the smallest building blocks known. Once the atom was the smallest building block known, and some scientists believed that they were indivisible. But there were 90 different chemical elements known and each element had a different kind of atom. Nobody understood why there were so many and how one kind of atom was related to another. Eventually scientists discovered that atoms are made of nuclei and electrons. But a great deal of research was required before these smaller building blocks and the true structure of the atom was discovered and completely understood.

The arrangement of all the chemical elements by Mendeleev into a periodic table was a major step in our understanding the nature of these elements and later to the understanding of the structure of their atoms. It was now clear that the different kinds of atoms followed a regular pattern which gave clues to the underlying structure. The first indications that Mendeleev's table would lead us to better understanding was its prediction of the existence of new chemical elements which had not yet been discovered. There were gaps in the table, and scientists immediately began searching for new elements which would fill in these gaps. Striking confirmations of the validity of this table were provided by the discovery of new elements which just filled gaps in the table.

Once the structure of the atom and its building blocks were understood, more and more of these building blocks were discovered, and the same problems kept arising at tinier and tinier levels with smaller and smaller building blocks. Each time scientists have searched for and found ways to arrange the known building blocks into tables exhibiting regular patterns which might give clues to their underlying structure. If there are gaps in the table, it is natural to predict the existence of new elements to confirm the systematics indicated in the table. When these new elements are found, the table becomes accepted and physicists try to understand what lies behind it.

The theory of atomic structure explained Mendeleev's table. Every element was made of atoms, and each atom had a nucleus and a number of electrons around it. The place of each element in Mendeleev's table was determined by the number of electrons in the atom. The electron has a negative electric charge. The nucleus has a positive charge which exactly cancels the total negative charge of all the electrons in the atom, so that the atom was electrically neutral. The simplest atom is the hydrogen atom, which has one electron and a nucleus with a positive charge equal to that of one electron. The biggest atom found naturally in nature is uranium, which has 92 electrons and a nucleus with a positive charge 92 times larger than the charge of one electron. All the 92 chemical elements having all possible numbers of electrons between 1 and 92 are now known; the missing ones in Mendeleev's table have all been discovered. There are now many new "transuranium" elements with more than 92 electrons that have been discovered in laboratory experiments which made them by banging two lighter nuclei together.

Next it was found that there were different kinds of atoms of the same element. They had the same number of electrons, but different weight. These were called isotopes. Hydrogen, the simplest element, was found to have three isotopes. In addition to ordinary hydrogen there is heavy hydrogen which is twice as heavy as ordinary hydrogen and combines with oxygen to make heavy water. There is also an even heavier hydrogen, called tritium, which is three times as heavy as ordinary hydrogen.

Soon there were so many known isotopes of each element that a new kind of table was made, a table of nuclei listing all the different isotopes of each element. Again there were gaps in the table and scientists searched for and found the missing isotopes. Investigations of the structure of nuclei by the sledgehammer technique soon revealed the building blocks of nuclei and explained the table of isotopes. All nuclei are made of two kinds of basic building blocks, which are called neutrons and protons. The proton is the nucleus of the lightest hydrogen atom. It has an electric charge exactly equal and opposite to the charge of the electron, but it is about two thousand times heavier than the electron. Thus the hydrogen atom which has one proton and one electron is electrically neutral. The neutron is very much like the proton, also about two thousand times heavier than the electron, but it is electrically neutral and has no electric charge.

The nucleus of heavy hydrogen, called the deuteron, contains one proton and one neutron. This makes it about twice as heavy as the proton, but with the same electric charge. The nucleus of the third isotope of hydrogen which is three times as heavy has one proton and two neutrons. All the isotopes of hydrogen have the same electric charge and are exactly neutralized by one electron.

All other known nuclei are built the same way, they contain protons and neutrons. The electric charge of the nucleus is just equal to the number of protons. Nuclei of different isotopes of the same element have the same number of protons and different numbers of neutrons. Names like carbon-14 and carbon-12 or uranium-235 and uranium-238 are used to distinguish between different isotopes of the same element. The common carbon nucleus has six protons and six neutrons and is twelve times heavier than the proton. It is called carbon-12. Carbon-14 is a rare isotope of carbon and has a nucleus with six protons and eight neutrons and is fourteen times heavier than the proton.

The neutron, proton and electron all behave like tiny magnets as well as carrying electric charge, and they are all spinning like a top. It is possible to measure the mechanical spinning motion, called angular momentum, of the proton, neutron and electron, and they are found to all have the same quantity of spin, which never changes. We can change the direction of a proton, electron or neutron spin by putting it in a magnetic field, but the magnitude of the spin always remains the same. We know that electric currents produce electromagnets, and that particles which have electric charge and are spinning like a top should behave like magnets. The electron behaves exactly in this way, and it is possible to compute the strength of its magnet very precisely by a theory called quantum electrodynamics.

Nuclear research continues from this point in two directions. One, called nuclear physics or nuclear structure physics, studies how the protons and neutrons are arranged in the nucleus and what the forces that hold them together are. The other, which is now called particle physics, studies the structure of the proton and the neutron. At first the proton and neutron were considered to be the final elementary stage in the investigation of the structure of matter, but now we know that they are made of even smaller building blocks.

The spinning motion of the particles and their behavior like tiny magnets provided important clues to the understanding of the structure of these tiny objects. When neutrons and protons are put together to make a nucleus, their spins combine to make the total spin of the nucleus. The deuteron, the heavy hydrogen nucleus, contains a neutron and a proton with their spins in the same direction, so that the spin of the deuteron is twice as big as the spin of a neutron or proton. The nucleus of normal helium, which is also called an alpha particle, contains two neutrons spinning in opposite directions and two protons spinning in opposite directions, so that all their spins cancel and the total spin is zero. The two proton magnets in the alpha particle also cancel one another and the same for the two neutron magnets, so that the alpha particle does not behave like a magnet. The total spin and the strength of the magnet for every nucleus can be measured in the laboratory, and the models for nuclear structure must explain how the spins and magnets of the neutrons and protons add up to give the observed value.

The strengths of the proton and neutron magnets gave the first indication that they are not elementary objects but have a more complicated structure. The proton has a magnet which is much stronger than that of a "spinning electrical top" in quantum electrodynamics. The neutron has no electric charge but behaves like a magnet made of spinning negative charge. This suggests that the neutron is not an elementary object with no electric charge but is made up of smaller building blocks having both positive and negative charges that exactly cancel one another. There can then be a net spin and a net magnetism because the negative charge is spinning differently from the positive charge, and therefore the magnetic effects of the positive and negative charges do not cancel. The strength of the neutron magnet was first measured in 1939 by two great experimental physicists, Luis Alvarez and Felix Bloch, who later won Nobel prizes for other very important discoveries. They passed beams of neutrons through magnetized iron and measured how the neutrons were affected by the magnetization of the iron. It was not until twenty-five years later that the neutron magnet was understood in terms of smaller building blocks with opposite charges.

Physicists attempted to study the structure of neutrons and protons and nuclei by bombarding them with high-speed particles from accelerators. To their great surprise, they discovered in their experiments a large number of new particles. Two families of new particles were found, called mesons and baryons. In addition there were a new kind of particles that physicists called "strange particles" when they were discovered, because of their peculiar behavior which was not understood. We now describe these particles as having a new kind of charge, different from electric charge, in addition to electric charge. Murray Gell-Mann called this new kind of charge "strangeness", showed that it was possible to define units of strangeness and demonstrated that strange particles were produced in pairs with equal and opposite values of strangeness, just like electric charge. A particle with two or three units of strangeness would be produced with two or three other strange particles each having one unit of opposite strangeness. The neutron and proton have zero strangeness, and the other particles which have been found have values of strangeness between $+3$ and -3.

The spins of all these particles were measured. Some had the same spin as the electron or proton and some of the first baryons discovered were found to have three times as much spin as the electron or proton. The first mesons discovered were found to have zero spin, and later other mesons were found which had exactly twice the spin of the electron or proton.

The particles came in families, called charge multiplets, like the neutron and proton, whose members were very similar. They all had the same spin and differed only in their electric charge. Some families were doublets, like the neutron and proton; others were triplets with electric charges $+1, 0$ and -1. There were also singlets with charge 0, and later some quartets were found. The families were generally denoted by Greek letters, like the Sigma (Σ) triplet, with the electric charge indicated by a superscript; e.g. Σ^-, Σ^o and Σ^+.

Soon there was need for another table to classify these particles. Gell-Mann and Yuval Ne'eman found that they fit very nicely into larger families of eight, called octets, which contained four charge multiplets, two doublets, a triplet and a singlet, all with the same spin.

Table 1 shows three such octets, a baryon octet with the same spin as the neutron and proton and two meson octets, one with no spin and one with double the spin of the proton. The particles are listed according to the values of their electric charge and strangeness. Table 1A shows the baryon octet which contains the neutron and proton and six other particles which are called the Sigma (Σ) triplet, the Lambda (Λ) singlet and the Xi (Ξ) doublet.

Table 1B shows the zero-spin meson octet which contains particles denoted by the letter K and the lower case Greek letters pi (π) and eta (η).

Table 1C shows the "double-spin" meson octet which contains particles denoted by the letter K^* and the lower case Greek letters rho (ρ) and omega (ω).

Table 1A.   The Baryon Octet

Strangeness	Charge = −1	Charge = 0	Charge = +1
0		neutron	proton
−1	Σ^-	Σ^o, Λ	Σ^+
−2	Ξ^-	Ξ^o	

Table 1B. The Zero-Spin Meson Octet

Strangeness	Charge = −1	Charge = 0	Charge = +1
+1		K^o	K^+
0	π^-	π^0, η	π^+
−1	K^-	$\overline{K^o}$	

Table 1C. The "Double-Spin" Meson Octet

Strangeness	Charge = −1	Charge = 0	Charge = +1
+1		K^{*o}	K^{*+}
0	ρ^-	ρ^0, ω	ρ^+
−1	K^{*-}	$\overline{K^{*o}}$	

The striking similarity between these three octets is exhibited by writing them together as

$$
\begin{array}{ccc}
\begin{array}{cc} n & p \\ \Sigma^- & \Sigma^o, \Lambda \quad \Sigma^* \\ \Xi^- & \Xi^0 \end{array}
&
\begin{array}{cc} K^0 & K^+ \\ \pi^- & \pi^0, \eta \quad \pi^+ \\ K^- & \overline{K^0} \end{array}
&
\begin{array}{cc} K^{*0} & K^{*+} \\ \rho^- & \rho^0, \omega \quad \rho^+ \\ K^{*-} & \overline{K^{*0}} \end{array}
\end{array}
$$

where n and p stand for neutron and proton respectively.

When Gell-Mann and Ne'eman first proposed their octet model, which they called "The Eightfold Way", the η meson had not yet been discovered and there were only seven known mesons in Table 1B. There were even some theorists who proposed a different model in which mesons were in septets. Again there was a Table like that of Mendeleev with a gap in it, and the discovery of the η filled the gap and confirmed the octet model.

After the first three octets were found with the same spin as the neutron and proton, no spin and double spin, the next group of particles found were baryons with triple the spin of the proton. But these did not fit into an octet. There were ten particles, a quartet

Table 2. The Triple-Spin Baryon Decuplet

Strangeness	Charge = −1	Charge = 0	Charge = +1	Charge = +2
0	Δ^-	Δ^o	Δ^+	Δ^{++}
−1	Σ^{*-}	Σ^{*o},	Σ^{*+}	
−2	Ξ^{*-}	Ξ^{*o}		
−3	Ω^-			

denoted by Delta (Δ), a triplet denoted by Σ^*, a doublet denoted by Ξ^* and a singlet denoted by Ω, as shown in Table 2.

In 1962, when this family was first proposed by Sheldon Glashow and J. J. Sakurai, the Ω^- was missing from the table, and these and other physicists suggested that experimenters should search for it. Other theorists who disagreed with this "Eightfold Way" said that it was all nonsense and that the Ω^- would never be found. When the Ω^- was found in 1964, the gap in this new Mendeleev table was filled and the Eightfold Way was accepted.

The next step was to understand why there were all these particles that fit so neatly into these octets and decuplets. There were also puzzling differences between the octets and decuplet that remained to be understood. The meson and baryon octets were not quite the same. The strangeness values in the meson octet are +1, 0 and −1; the strangeness values in the baryon octet are 0, −1 and 2. There was no obvious reason why the triple-spin baryons should be in a decuplet rather than an octet. Soon after the discovery of the Ω^-, Murray Gell-Mann and George Zweig showed that everything could be explained simply if all these particles were made from smaller building blocks which Gell-Mann called quarks.

Why Particles Seem to be Made out of Quarks

To see how our observations of the properties of all these particles gives the impression that they are made out of quarks, let us first consider one simple property, the electric charge. The simplest constituents of the nucleus, the neutron and the proton, are very similar to one another except for their electric charge. The proton has a positive charge exactly equal to the charge of the electron. The neutron has no electric charge at all. There is a family of particles called baryons, which include the neutron and the proton.

Some of these baryons have charges zero and $+1$ like the neutron and proton, some of them have charge -1, and some have charge $+2$. There are no baryons which have values of electric charge different from $-1, 0, +1$ and $+2$. This seems very peculiar. There is no simple explanation, other than a quark structure, for why these and only these values of electric charge are found.

However, if we assume that all baryons are made of three quarks, we get exactly these values of electric charge. There are two kinds of quarks, one with electric charge $+2/3$ that of the proton and one with a charge $-1/3$. We then see that:

Three quarks with a charge of $+2/3$ have a total charge of $+2$

Three quarks with a charge of $-1/3$ have a total charge of -1

Two quarks with a charge of $+2/3$ each and one with $-1/3$ have a total charge of $+1$

Two quarks with a charge of $-1/3$ each and one with $+2/3$ have a total charge of 0.

These are all the charges that can be made with three quarks, and they just fit the values observed experimentally for the electric charges of the family of particles called baryons.

But have we really explained why the baryons have their peculiar charge values? It looks like we have simply confused things by raising a new question of why the quarks have the charges $+2/3$ and $-1/3$. Have we really made progress in our understanding of the electric charges of the baryons by this quark model, or are we back where we started? If all we knew about baryons were their electric charges, we would indeed have not learned very much from this quark model. But baryons also have other properties that are measured in experiment, and we can ask whether the quark structure explains these as well. It turns out that many other properties are also explained.

In addition to the normal particles, like the neutron and proton, there are also the strange particles. The neutron and proton have zero strangeness, and the other baryons which have been found have values of strangeness between zero and −3. The celebrated omega-minus particle has strangeness equal to −3. But no particles in any baryon family have values of strangeness different from 0, −1, −2, and −3. This is explained in the quark model by saying that there is a strange quark in addition to the normal quarks which make up the neutron and proton. This strange quark has a value of strangeness of −1. The baryons are made of three quarks, and can have values of strangeness of 0, −1, −2, and −3, depending upon whether they contain no strange quarks, one strange quark, two strange quarks or three strange quarks. But we can say even more about the strange particles. There is only one kind of strange quark, and it has electric charge of −1/3. From this we see immediately that particles which have strangeness −3 like the omega-minus must have an electric charge of $3 \times (-1/3)$ and cannot have any other value. A particle which has strangeness −2 has two strange quarks with a charge of $2 \times (-1/3) = -2/3$ and a third normal quark which can have a charge of either −1/3 or +2/3. Particles of strangeness −2 are then allowed to have electric charges of either −1 or zero. In the same way, a particle with strangeness −1 is allowed to have electric charges of −1, 0 and +1. So the quark model gives us complicated rules about which values of electric charges are allowed for strange particles.

We can now describe these rules more precisely. The normal quarks found in the neutron and proton have been given the names "up" and "down", and are denoted by the letters u and d. The up quark has electric charge +2/3; the down quark has electric charge −1/3. The strange quark, denoted by the letter s, has electric charge −1/3. Table 3A shows a listing of all possible combinations that can be made from three u, d and s quarks arranged in different ways, and gives their total electric charge and strangeness. Experiments show that exactly these baryons and no others are found.

Table 3A.   Baryons with Triple Spin Made From Three Quarks

Baryon	Charge = −1	Charge = 0	Charge = +1	Charge = +2	Strangeness
Δ	ddd	udd	uud	uuu	0
Σ*	dds	uds	uus		−1
Ξ*	dss	uss			−2
Ω	sss				−3

This is just the baryon decuplet that we have seen in Table 2. The quark model also gives simple explanations for the octets shown in Table 1. The quark has a spin which is equal to the spin of the neutron and proton. The particles in the baryon decuplet are all made of three quarks with all of their spins lined up parallel, so that the total spin is three times the spin of the quark. But it is also possible to line up the quark spins so that two are parallel and the third is opposite to the other two. The total spin is then just equal to the spin of one quark, because two equal and opposite spins cancel one another. This is the way the neutron and proton are constructed from quarks.

But why are there only eight particles in the family of the neutron and not ten? The answer is that quarks of the same kind like to have their spins parallel. This comes out of the more sophisticated version of the quark theory in a very natural way, but it is too complicated to explain here. So a particle with three identical quarks, either uuu, ddd or sss, must have all its quark spins parallel and have a total spin three times that of the proton. So the particles with all three quarks identical are absent from the octet containing the neutron and proton. The proton (uud) has two identical u-quarks with their spins parallel and the spin of the odd third quark in the opposite direction. The same is true for all those other baryons in this octet which have two identical quarks. But a baryon in which no two quarks are alike has an additional possibility. There are two different particles in the baryon octet made from one *u*, one *d* and one *s* quark. In the Λ the spins of the *u* and *d* quark are opposite. In the Σ^o the spins of the *u*

Table 3B. Baryons with Single Spin Made from Three Quarks

Baryon	Charge = −1	Charge = 0	Charge = +1	Charge = +2	Strangeness
Nucleon		$u(dd)$	$(uu)d$		0
Λ		$[ud]s$			−1
Σ	$(dd)s$	$(ud)s$	$(uu)s$		−1
Ξ	$d(ss)$	$u(ss)$			−2

and d quark are parallel and the spin of the s quark is opposite to the other two. Thus the quark model explains in a natural way why there is a baryon octet with a total spin equal to the spin of a quark and a baryon decuplet with three times the spin of a quark.

Table 3B shows a listing of all possible combinations that can be made from three u, d and s quarks arranged in different ways with two spins opposite, with the restriction that quarks of the same kind must have their spins parallel. These correspond exactly to the baryon octet of Table 1a. We have used the name nucleon to denote either the neutron or the proton. The spin orientations are shown by putting parentheses as in (uu)s around a pair of quarks whose spins are parallel and square brackets as in [ud]s around a pair of quarks whose spins are opposite.

One of the first successes of the model was the explanation of the magnetic properties of the baryons. The strengths of the magnets in the neutron and the proton were computed by adding up the strengths of the three quark magnets in each. We find that the magnets of the neutron and proton must point in opposite directions, and that the strength of the neutron magnet must be exactly 2/3 the strength of the proton magnet. This result is in striking agreement with the experimentally measured values of the magnetic properties of the neutron and proton.

We can see in a crude way how this computation was carried out. The main point is that the u and d quarks have opposite signs of charge, so that their magnets point in *opposite* directions they are spinning in the *same* direction. But in the proton, where the *u* and *d* quarks are spinning in *opposite* directions, the magnets are all pointing in the *same* direction and add up to give a strong magnet. The same is true in the neutron.

Let us assume that the strength of each quark magnet is proportional to its electric charge. In the proton the two *u* quarks with charge $+2/3$ each contribute $+2/3$ units of magnetism, while the *d* quark with charge $-1/3$ is spinning in the opposite direction and contributes $+1/3$ unit. The total strength of the proton magnet is thus $+5/3$ units. In the neutron, the two *d* quarks with charge $-2/3$ each contribute $-1/3$ units of magnetism, while the *u* quark with charge $+2/3$ is spinning in the opposite direction and contributes $-2/3$ units. The total strength of the neutron magnet is thus $-4/3$ units. We thus see that even though the neutron has zero charge, the strength of its magnet comes out to be large and negative. Our crude calculation thus gives the result that the magnets of the neutron and proton must point in opposite directions, and that the strength of the neutron magnet is 4/5 the strength of the proton magnet. This approximate result is not far from the correct ratio 2/3 obtained from the exact calculation using the quantum-mechanical theory of angular momentum.

This is typical of the way the quark model is used. We began with fitting the properties of the quarks to get the electric charges of the baryons to come out right. Then we found that we also got the strangeness and spins of the baryon to come out right, and we also got a very precise correct value for the ratio of the magnetic strengths of the neutron and proton. This is why we say that the particles behave "as if they were made out of quarks". Their electricity, magnetism and spin would be very hard to understand if they were not made up out of other building blocks. We would not understand, for example, why the neutron which has no electric charge behaves like a magnet similar to the proton which has electric charge. But by assuming that the baryons are made out of the quarks which have certain definite values of electric charge, everything else falls into place.

The mesons have a different structure. To understand them we must introduce the idea of antiparticles. Each particle, like the proton or electron, has an antiparticle, the antiproton or antielectron, also called the positron. Corresponding particles and antiparticles have exactly the opposite properties. The electric charge of the antiproton is negative, opposite to that of the proton; the electric charge of the positron is positive, hence the name, opposite to the negative charge of the electron. The antiparticles of strange particles have the opposite values of strangeness. All the antiparticles of the baryons in the octet and decuplet of Tables 1A and 2 are known, and they fit into a different octet and decuplet.

The antiparticles of the mesons are already in the meson octets of Tables 1b and 1c. The π^+ is the antiparticle of the π^-, etc. The π^o, η, ω and ρ^o which are neutral and have no electric charge are their own antiparticles. The $\overline{K^o}$ is the antiparticle of the K^o particle, and it is common to use a bar over a symbol for a particle to denote its antiparticle.

The quarks also have their antiparticles, denoted by $\bar{u}$, $\bar{d}$ and $\bar{s}$. Their electric charges are $(-2/3, +1/3, +1/3)$ just opposite to the charges of their corresponding quarks, and the strangeness of the $\bar{s}$ antiquark is $+1$.

Mesons are constructed from a quark and an antiquark. If the spins of the two are opposite, the meson has zero spin; if the spins are parallel, the meson has double the spin of the proton. This leads naturally to the two kinds of mesons in Table 1B and 1C, with no spin and double spin. We can also see that the right values of electric charge and strangeness come out, as shown in Table 4.

Table 4.   Mesons Made From a Quark and an Antiquark

Meson	Charge = -1	Charge = 0	Charge = $+1$	Strangeness
K		$d\bar{s}$	$u\bar{s}$	$+1$
π, η	$d\bar{u}$	$d\bar{d}, u\bar{u}, s\bar{s}$	$u\bar{d}$	0
$\overline{K}$	$s\bar{d}$	$s\bar{u}$		-1

Here we see exactly the same electric charge and strangeness as in the meson octets, but there are nine particles instead of eight. There are three particles in the center of the table with zero charge and zero strangeness, instead of only two. In fact these extra ninth members of the meson "nonets" have been found. The zero-spin nonet has a ninth member, called the η' and the double-spin mesons have a ninth member called the ϕ. The discoveries of these ninth mesons gave further confirmation to the quark model.

Gaps in the Next Mendeleev Table

At this point we have the u, d and s quarks which are inside the nucleus and its constituent neutrons and protons. But we also have the electron moving around outside the nucleus. Experiments on nuclear beta decay discovered another particle emitted from nuclei together with electrons. This particle, which is called the neutrino (ν), is an electrically neutral partner of the electron, just like the neutron is an electrically neutral partner of the proton. The neutrino has no charge, like the neutron, but it is very light, like the electron. The neutrino mass is much smaller than the electron mass which is about 1/2000-th of the mass of the proton. In fact the neutrino mass is so tiny that it has not yet been measured.

After the neutrino was established, everyone thought that the neutron, proton, electron and neutrino were all the particles needed to describe the structure of matter, and that the nuclear physicists had achieved their dream of finding the new four ultimate elementary constituents of nature, the successors to earth, air, fire and water.

This dream was shattered by the discovery of a new heavier electron called the muon (μ). It behaved exactly like the electron and had the same electric charge, but was just heavier. Nobody needed this new particle, and the attitude of the physics community to the discovery of the muon was summed up in Prof. I. I. Rabi's famous remark: "Who ordered that?" Soon many other particles were found and it was clear that the neutron and proton were not elementary.

The neutron, proton, and all the mesons and baryons are now known to be made of the smaller quark building blocks.

A neutrino partner of the muon was also found. The two neutrinos are now called "electron neutrino" (v_e) and "muon neutrino" (v_μ) to distinguish them from one another. These particles which do not remain inside the nucleus and do not feel the strong forces binding quarks together are called leptons. Quarks and leptons were considered to be completely different objects until J. D. Bjorken and Sheldon Glashow put them together in a new Mendeleev table.

But the search for more building blocks continued, even though some physicists believed that now we have found them all.

Excitement About New Particles

In November 1974 particle physicists had a great surprise. A single issue of the well known physics journal, *Physical Review Letters*, contained three communications, one from a group of 14 physicists from M.I.T. and Brookhaven National Laboratory on the American East Coast, one from a group of 35 at the Stanford Linear Accelerator Laboratory on the West Cast and one from a group of 43 at the Italian Frascati Laboratory near Rome. These three communications reported the discovery and properties of a new particle whose existence had been completely unexpected. A second new particle was discovered shortly afterwards and the world of high energy physics was thrown into total confusion. The journals were soon flooded by papers proposing theoretical interpretations of the new particles. One journal, *Nuclear Physics*, adopted an editorial policy of holding all these articles and not agreeing to publish any of them until the dust settled a bit. New experiments planned for the next few weeks might well render 99% of the papers obsolete.

The discovery of the new particles is now called the "November Revolution". It was reported in the *New York Times* and other newspapers and news magazines. Some of you may have seen them and wondered what it is all about. It may seem peculiar that a weekly journal like *Physical Review Letters* should carry three reports of the same discovery made independently on the East Coast, on the West Coast and in Italy all in the same week. Papers with 35 authors must also appear peculiar to those not familiar with the ways of High Energy Physics. So let me explain how research in experimental high energy physics is done these days.

In the summer of 1974 particle physicists were very confused. It had been predicted that the "strange" quark must also have a partner, which was called the "charmed" quark. But the charmed quark had not yet been discovered, and most physicists did not believe that it existed at all. Experiments with high energy collisions between protons produced a complicated mess of many particles which nobody could unravel. Then in November 1974 came the discovery of a new particle, the psi, simultaneously at the Brookhaven National Lababoratory in New York and at the Stanford Linear Accelerator Laboratory in California. Today physicists speak of the "November revolution", which led to the discovery of the charmed quark and the confirmation of the new model with two generations of families of four particles. The psi is now known to be a compound of two charmed quarks bound together by strong forces. But it took a year before this was conclusively proved. Physicists argued back and forth about this new psi particle. Some said that the charmed quark had been discovered; others said that the psi could not possibly contain charmed quarks. Further experiments on the psi gave peculiar results; some agreed with the charm hypothesis, others did not.

Meanwhile Professor Martin Perl, working with the group investigating the psi at Stanford also pushed his own idea, a search for a new heavy electron, heavier than the muon. There was no theoretical reason to look for such a particle, and most physicists thought that the idea was completely crazy. But Perl continued his experiments, and by the end of the summer of 1975 there seemed to be evidence for such a new particle, which he named the "tau". But this was still highly controversial.

During the summer of 1975 there were a number of international conferences on particle physics in Europe and the United States which featured new ideas on the nature of the new particles. At the beginning of the summer, the experimental evidence for the charm hypothesis did not look good, and the experts were betting on other explanations. By September, new results favored charm, and there were indications that most of the confusion arose because Perl's new heavy electron happened to be produced in the same experiments that produced the charmed quarks, and it was difficult to unscramble the two effects. The psi particle and the tau particle were never seen directly in experiments because they lived only for a fraction of a billionth of a second and then decayed into several other particles. Only the particles produced in their decays were observed and measured. The analysis of the experiment involved figuring out how some of the observed particles came from the decays of other particles which lived for too short a time to be observed directly.

The charmed quarks in the psi particle were called "hidden charm" because it was harder to prove that they were really there when there were two of them. The search was on for particles that contained only a single charmed quark. This was called "naked" charm, and was found in the fall of 1975 by a group working on the Stanford experiments.

At the same time that charm was finally confirmed, and the two-generation "standard model" seemed to explain everything, Perl's tau particle came along and upset it all again. This new heavy electron was greeted in the same way that I. I. Rabi greeted the discovery of the first heavy electron, now called the muon. "Who ordered that?" As soon as a discovery is made which fits together all the pieces of the puzzle, someone always finds a new piece which doesn't fit anywhere, and the puzzle starts all over again. Here Haim Harari pointed our that the tau indicated the presence of a new "generation" of particles including two new quarks which Harari called "top" and "bottom".

In 1977 a new particle called the upsilon was discovered by Professor Lederman's experimental group working at the Fermi National

Accelerator Laboratory (Fermilab) in Illinois. This was immediately shown to be made from a new kind of quark, Harari's "bottom" quark. This new quark seemed to belong together with Perl's new heavy electron in another family, and the "three-generation" model was born. There was already evidence for a neutrino-like partner to Perl's tau particle, and the search was on for Harari's "top" quark. But there was a controversy among the physicists about naming the new quark family. Lederman's quark was called the "b" quark and its undiscovered partner was called the "t" quark. But physicists who liked simple names like "up" and "down" said that "t" and "b" should be called by Harari's names "top" and "bottom", while those who liked picturesque names like charm and strange called them truth and beauty.

The "b" quark was "hidden" in Lederman's upsilon particle, which contained two of them, but the particles with a single "naked" b quark were soon found. The t quark has now been found as well, after the particle physicists had a great time describing their results and theories with the new fancy names. Physicists had discovered the naked beauty and then were searching for truth. But until they found the top or truth quark some theorists speculated that perhaps there was no top quark and developed "topless models with hidden charm and naked bottoms".

The discoveries of Lederman and Perl revealed the existence of a new third generation of fundamental particles, and opened up a new era in particle physics research which is still going on. The properties of these new particles are being studied, and the top quark to finish the family has now been found. But there is still no good theory of why there are three identical generations of particles, and nobody knows whether it stops here or if there are more generations.

Once again there are too many particles and the theorists are trying to explain why there are so many. There are two competing approaches. One considers them as compounds, made of a much smaller number of even tinier objects not yet discovered. The large number of chemical elements and later the large number of different nuclei were explained by showing them to be compounds. The other approach looks for a unified description of all the different forces of nature, like the unified field theory that Einstein tried to find in his later years. Electricity, magnetism, radio waves, microwaves, light waves, infrared radiation, ultraviolet radiation, nuclear gamma radiation and X-rays have been unified in this way. They are now all described as radiation of the same kind of particles called photons. The recently discovered W particle brings the weak force of radioactivity into this unified family of forces. Perhaps all the generations of particles are all different aspects of the same particle, like all the photons. Nobody knows the answers to these questions. That's what makes the puzzle so interesting.

Why these New Particles were so Confusing — A Historical Survey

In the 1940's particle physicists called themselves "elementary particle physicists" under the mistaken impression that the particles they were studying were the elementary constituents of matter. In the past forty years so many new particles had been discovered that one thing was now clear. None of them were elementary. Perhaps this was a new kind of atomic spectroscopy, a new set of energy levels or energy states which reflect some simple underlying dynamics, just as the set of states of the hydrogen atom reflect the dynamics of the proton and electron inside. Particle physicists were making some headway in the interpretation of the large number of particles by techniques similar to those used in the early days of atomic physics. Particles were classified into something like a periodic table. Then they tried to understand this periodic table in terms of models of possible constituents and their motion. The main trouble is that these constituents have not yet been discovered experimentally. These "elementary building blocks of matter" are sometimes called "quarks". Many people have looked for them, but nobody has found even one isolated quark.

When experiments were performed at higher and higher energies, higher and higher excited states of particles are found, just as in atomic physics. These excited states decay very rapidly to the lower states. The higher up you go, the faster they decay. The new particles just discovered have a very high excitation energy and they do in fact decay to the lowest known states. But they decay much more slowly than they have any right to. It is because they decay so slowly that they are considered so peculiar and have been missed all this time. No one in his right mind would submit a proposal to search for a particle with an excitation energy of three times the mass of the proton and a lifetime ten thousand times longer than the lifetimes of other particles known at much lower excitation.

The new particles were not discovered simultaneously in three different places during the same week. Peculiar effects had been previously noted both on the East Coast and on the West Coast in experiments performed for entirely different reasons. These were not taken too seriously. A careful experimenter who sees such effects first checks carefully for something wrong with the apparatus or bugs in the computer program which processes the data. He would not immediately claim an exciting new discovery or even ask an accelerator program committee for more machine time to look into the effect, as this would mean taking some other group off the machine. But as soon as the groups in two different laboratories realized that they had independently observed the same kinds of peculiar effects using completely different accelerators, completely different experimental apparatus and completely different data processing systems, it became clear that this was a real effect due to the particles themselves and not a fake effect due to something wrong with the apparatus. It could only be some new kind of particle with unsuspected properties. Fortunately, high energy physicists working in different laboratories talk to one another and do not keep their results secret. This cross communication leads to more efficient research and more rapid progress. Thus each American group heard about the results of the other, and the Italian group was also informed. At this point all groups began working feverishly to get as much information as possible about the particle and its properties, and the first letters reporting the results were submitted for publication the same week.

At the Weizmann Institute we were fortunate to have one of our professors on sabbatical at the Stanford Linear Accelerator Laboratory that year. He kept us posted on the new information with regular letters and occasional cables, beginning with a telegram on the day of the first discovery. At first it was a bit frustrating to get letters a week after they were written and to wonder what had happened during that week. We thought of telephoning but were assured that it would not be practical because there were so many wild rumors of preliminary results floating around. It would take at least thirty minutes on the phone every day to keep us up to date, and most of the time would be spent explaining why many of the things we heard yesterday were no longer believed to be right. So we relaxed and settled for week-old news bulletins on the new particles.

When I first heard about the new discovery, I tried to explain the structure of the new particles with a model that I had used in previous work. The model accounted for many of the observed properties of the particle, but had some serious difficulties as well. I wrote to Stanford about it and explained some ideas for further work which might avoid the difficulties. By return mail came the answer that the theoretical group at Stanford had already noted this, and that a group of about fifteen theorists led by the director of the group had made an exhaustive analysis of this particular model, had investigated fifty different ways of modifying it to overcome the difficulties, including the particular ones I had suggested, and had found that nothing worked. I therefore decided that there was no point in competing with this juggernaut, and that we would just have to wait here and digest the information as it came. There was no sense in trying anything else unless we were sure that we had a radically new idea which had not already been picked to pieces in fifty versions by fifteen theorists.

The excitement and chaos around the discovery of the new particles had one very clear moral. The best brains in theoretical physics including a number of Nobel Prize winners had been very active in this field and none of them predicted that such particles should be observed or told the experimenters how and where to look for them. For a year after the discovery there was complete confusion among the theorists who could not agree on any satisfactory explanation for these particles. Perhaps this should give us a feeling of humility. I recall the words my father used to say when I was a small boy and thought that I had been very clever: "If you would know what you don't know, you would know more than you know."

The New Mendeleev Table including all Three Generations

The discovery of Perl's new lepton called the "tau" (τ) and its neutrino partner called the "tau-neutrino" (ν_τ), together with the new t and b quarks, completed a third "generation" of quarks and leptons. The quark–lepton Mendeleev table now has a new line.

So far we have been discussing the search for the tiniest building blocks of matter and reached the level of quarks and leptons. The other part of the story is the understanding of the forces or cement that hold these constituents together. We now turn to this aspect of the story of the structure of matter.

Table 5B. Quarks and Leptons

Quarks		Leptons	
Charge $= +2/3$	*Charge* $= -1/3$	*Charge* $= 0$	*Charge* $= -1$
u	d	ν_e	electron
c	s	ν_μ	muon (μ)
t	b	ν_τ	tau(τ)

6

The Forces of Nature

Gravity and Electromagnetism

In our everyday experience we encounter many forces which push and pull matter in different directions. One is the force of gravity, which makes everything fall to the earth. Newton discovered that this same force of gravity acted between all bodies of matter and explained the motion of the Earth and the other planets around the Sun and of the Moon around the Earth. Using his laws of gravitation and motion he was able to calculate the orbits of the planets and the moon with great precision.

Since Newton's time astronomers have made more precise observations of these orbits and discovered new planets, while Newton's calculations have been refined to describe finer details of these orbits. In fact the existence of the planet Neptune was predicted by such calculations. The orbit of the planet Uranus as measured by the astronomers did not exactly fit the orbit calculated using Newton's laws and the gravitational forces on Uranus from the Sun and all the known other planets. The scientists decided that there must be some additional force acting on Uranus and found that the deviations in the orbit of Uranus could be explained if there was another new planet in an orbit farther away from the Sun. Their calculations showed where this new planet must be and the astronomers looked and discovered Neptune. Today Newton's laws are used to describe the motion of rockets and other projectiles shot from the earth and the motion of artificial satellites launched by man into orbits around the earth.

This approach of Newton to the forces of nature has been carried on by physicists to describe other forces. The twin aims of this approach are to unify forces that seem to be different by showing that they have a common origin and description, and then to develop the means of calculating precisely how these forces act and how matter behaves under the influence of these forces.

The next great step forward in our understanding of forces was the unification of electricity and magnetism by the British physicists Faraday and Maxwell. Faraday's experiments showed that electricity and magnetism which had previously been believed to be very different types of forces were in fact intimately related. His discovery that magnetic forces could generate electric currents and his formulation of the laws of "electromagnetic induction" laid the basis for the modern electric motors and generators which are so important in our everyday life today. In 1864 Faraday's results were described by Maxwell in a unified theory called electromagnetism originally developed to explain electricity and magnetism but which has turned out to explain very much more. We know now that Maxwell's theory explains all electrical, magnetic, chemical and optical phenomena.

Maxwell's theory also predicted the existence of radio waves before their discovery by showing how the energy of an electric current in the antenna of a radio transmitter could be sent through space to generate a current in a receiving antenna far away. Maxwell's theory showed that the Sun's rays are also electromagnetic radiation, generated by billions upon billions of antennas in the Sun. They send their energy through space to tiny antennas in our bodies which make us feel the energy as heat and to tiny antennas in our eyes which enable us to see light. Lasers, microwave cookers, x-ray photographs and a host of other phenomena are all described by Maxwell's electromagnetic theory with tiny antennas deep inside an atom sending energy to receiving antennas in other atoms.

In 1905 Einstein showed that Maxwell's electromagnetic radiation consisted of tiny particles called photons which carried the energy from the transmitter to the receiver. Einstein believed that all forces of nature were linked together. He spent the last years of his life in an unsuccessful search for a "unified field theory" more general than electromagnetism that would also include gravitation and describe all these forces.

The discovery that the atom consists of a nucleus with positive electric charge and electrons with negative electric charge moving in orbits around the nucleus led to even further applications of Maxwell's electromagnetic theory. All the forces within the atom and between one atom and another were shown to be explained by electromagnetism. This included the forces that kept the electron moving around the proton in the hydrogen atom and the forces between two hydrogen atoms and an oxygen atom to bind them together to make a water molecule. All of chemistry was suddenly revealed to be explainable in terms of Maxwell's electric forces. The precise description of atomic phenomena required a new generalization of Newton's laws of motion which is now called quantum mechanics. A detailed description of the quantum theory is beyond the scope of our present treatment. Our pictures of the atom as a miniature solar system are not strictly correct. But these pictures are adequate for an understanding of many properties of the atom, and we will not delve here into the mysteries and revolutionary ideas of the quantum theory.

Forces and Energy in the Nucleus

The force that holds the electrons and the nucleus together in the atom is ordinary electricity. The nucleus has a positive electric charge, the electron has a negative electric charge, and particles with opposite electric charge attract one another. All atomic phenomena were successfully described using the model with electrons moving in orbits around the nucleus and subject only to electrical forces. No other forces were needed to explain the behavior of atoms.

Nuclei are made of neutrons and protons. The neutron has no electric charge. The protons all have positive electric charge and repel one another. Electric forces alone would break up a nucleus into individual protons moving away from one another and have no effect on the neutrons. There must be another kind of force different from electricity which holds the protons and neutrons together.

By the middle of this century physicists were attempting to study the nature of this force by shooting beams of protons against against the protons in a hydrogen target. Again using the approach of Rutherford, they compared the results of their experiments with the orbits calculated for their protons if only electrical forces were acting. They found that there was no evidence for any new force until the protons came very close together. There was evidence for a new force only when the distance between two protons was less than one millionth of one millionth of a centimeter, or 10^{-12} centimeters in the commonly used notation for small distances. The measurements showed that this force was very strong and was attractive.

Similar results were obtained in experiments shooting beams of neutrons against protons. Here there was no electrical force and there was no effect at all of the protons on the neutron beams until the neutrons came very close to the protons, again less than 10^{-12} centimeters. This new force could explain why protons and neutrons stuck together in the nucleus. Some physicists call this new force the "nuclear glue" that holds the nucleus together.

It has taken many years for physicists to unravel the mysteries of this very short range nuclear force, and it is still not fully understood. But many properties of the nucleus became clear once it was realized that there were two kinds of forces at work: the electricity which tries to push the protons apart, and the new nuclear force which holds the neutrons and protons together. When two nuclei are far apart, they only feel the repulsive electric force between them and will not come together. But if they are shot at one another with sufficient force to overcome the electrical repulsion and come within one millionth of one millionth of a centimeter of one another, the strong new force takes over and they are held together to make a bigger nucleus. This process is called fusion.

On the other hand, if we can hit a piece of a nucleus hard enough so that it moves far enough away from the rest of the nucleus, farther than one millionth of one millionth of a centimeter, it will no longer feel the strong new force attracting it to the rest of the nucleus. It will feel only the repulsive electric force and the two pieces will move apart. This splitting of the nucleus into two pieces is called fission.

We thus see that the processes of nuclear fusion and nuclear fission arise naturally in any picture where two nuclei always repel one another at large distances but are very strongly attracted when they are brought close enough together. This interplay of the two kinds of forces explains how nuclear energy can be generated.

Natural uranium found in mines contains two isotopes called uranium-235 and uranium-238. Each contains 92 protons. Uranium-235 contaains 143 neutrons; uranium-238 contains 146. In 1936 the German scientists Hahn and Strassman discovered nuclear fission when they found that they could split into uranium into barium which contains 36 protons and krypton which contains 56 protons. Their experiment could not determine which isotope of uranium had been split, nor which isotopes of barium and krypton were produced. Today we know that they had split the very rare uranium isotope, U-235, which only constitutes 0.7% of ordinary natural uranium, and that many different isotopes of barium and krypton can be produced in uranium fission, depending upon which way the neutrons go when the uranium nucleus splits. There are also many other ways that a uranium nucleus can split into two fragments; sometimes it is baryon and krypton, other times it is other nuclei.

How do we get energy from fission or fusion? Suppose we take a barium nucleus and a krypton nucleus and try to put them together to make uranium. We have to exert force against the electric repulsion between the two nuclei to bring them to the point where the attractive nuclear force can take over. Once we get "over this hump" the strong nuclear force will pull the two pieces together and make them into a uranium nucleus. We have to put in energy in order to get the two nuclei close together.

We may gain energy from the powerful nuclear force after we are over the hump. But do we get out more energy than we put in? That is the question.

This situation with two opposing forces is somewhat like the problem of pushing a wagon over the top of the hill. Until we reach the top the force of gravity is against us and tries to pull the wagon backwards. Once we are over the hump it is with us and tries to pull it forwards. We must put in energy to get the wagon over the top. We get energy back when it goes down hill. We can measure this energy balance if we drive the wagon up the hill with an electric motor connected to a battery. Once we reach the top we connect the wheels to an electric generator which recharges the battery. If we go down the hill far enough we will get back enough electrical energy to recharge the battery to the same level that we started. If we go down even further we will gain energy.

A more practical example of this energy balance is the hydroelectric energy gained from a waterfall. Gravity pulls the water over the waterfall, and the energy of the motion of the water can be used to turn the wheels of electric genrators. We can gain energy as long as the water falls from a higher level to a lower level. The Mediterranean Sea is at a higher level than the Dead Sea. Thus energy gan be gained by letting water fall from the Mediterranean into the Dead Sea. But there are mountains between the two seas which prevent the water from moving freely from one to another. If water from the Mediterranean is pumped up to the top of the mountains, it will then fall into the Dead Sea and give us energy. Because the Dead Sea is lower than the Mediterranean, we know that we can gain useful energy in this process. The energy we gain when the water falls from the mountains into the Dead Sea is greater than the energy we need to pump the water from the Mediterranean to the top of the mountains.

Water can also be pumped up to the top of the mountains from the Dead Sea and give us energy by falling into the Mediterranean.

In each case we must put in energy to run the pumps, and we get back energy from the waterfall. But because we know we cannot gain energy by moving water from a lower level to a higher level, we know that it must cost us more energy to pump water from the Dead Sea to the top of the mountains than we will get back by letting it fall into the Mediterranean.

The energy balance in nuclear fission and fusion is like the energy balance in moving water from one sea to another over a mountain or barrier. We first have to put energy in to get over the barrier; then we can get energy out. But whether we gain or lose energy in the end depends upon the energy levels of the two sides. We can gain energy in the long run if we move from a higher level to a lower level, even if we initially have to pump energy in to get over the barrier. But we can never gain energy by moving from a lower level to a higher level; we will always have to pump in more energy than we get out.

In the fission of uranium into baryon and krypton we must somehow pump enough energy into the nucleus to push the baryon and krypton nuclei inside far apart against the strong nuclear forces holding them together. But once they are far enough apart, farther than 10^{-12} centimeters, they no longer feel the strong nuclear force and they are pushed apart by the powerful electric repulsion. It is as if we have pumped in enough energy to push them over the mountain or barrier and they now coast down with increasing speed which can be turned into useful energy. In the case of uranium fission the electric repulsive force pushes the baryon and kryupton nuclei apart with high speed and these nuclei then collide with other atoms and nuclei and turn their energy of motion into heat energy. In an energy-producing nuclear reactor this heat energy is used to produce steam which then runs electric generators.

In the case of uranium, we gain energy from fission, not from fusion. The uranium nucleus is at a higher energy level than the baryon and krypton nuclei that are produced in fission. If we push baryon and krypton nuclei together against the electrical forces pushing them apart, we must pump in more energy than we will get back when the nuclear forces take over and pull the nucleus together. We cannot gain energy by putting baryon and krypton together to make uranium. We can gain energy by splitting the uranium into baryon and krypton.

There are other nuclei where we gain energy from fusion and not from fission. The nucleus of heavy hydrogen, called the deuteron, contains one proton and one neutron. By putting two deuterons together we can make the helium nucleus which contains two protons and two neutrons. Here again we must overcome the repulsive electric force between the two deuterons in order to bring them close enough to allow the strong nuclear force to bring them together to make a helium nucleus. But once we pump enough energy in to get the two deuterons "over the barrier", the strong nuclear force will take over. Here again we can also pump energy into a helium nucleus and break it up into two deuterons. But in this case the helium nucleus is at a lower energy level than two deuterons. If we can put two deuterons together to make helium, we can gain more energy than we pumped in to bring the deuterons together. If we break up helium into two deuterons, we must pump in more energy than we get out.

Uranium fission and hydrogen–helium fusion are the two basic processes from which nuclear energy is produced today. Both cases make use of the opposite natures of the long range repulsive electrical force between two nuclei and the very short range attractive nuclear force. We begin with a situation in which one force is dominant, pump energy in until we reach the top of the mountain, where both forces are equal, and gain back energy from the other force on the other side. In fusion we pump energy in against the electrical force and gain it back from the nuclear force. In fission we pump energy in against the nuclear force and get it back from the electrical force. In both cases we have to get over the top of the barrier where both forces are equal. Nuclear physicists call this barrier the Coulomb barrier, in honor of the French physicist Coulomb who discovered the law of electric force between electric charges.

For small nuclei like hydrogen and helium, energy is always gained by fusion and lost by fission. The energy of attraction by the nuclear force is always stronger than the energy pumped in against the electric force to bring the pieces together. For large nuclei like uranium the reverse is true, and this is easy to understand. The uranium nucleus is so big that the neutrons and protons on opposite sides are already too far apart to feel the strong attraction of the nuclear force. Nuclei a little bigger than uranium cannot stay together at all; they break up almost immediately because pieces on opposite sides are pushed apart by the electrical force and can never get close enough to be held together by the attractive nuclear force.

If we try to build up nuclei by putting more and more neutrons and protons together we start with a situation where the nuclei are small and we can gain energy by putting nuclei together. But eventually we reach a point where we do not gain energy any more because the nuclei are too big and we can't bring all the neutrons and protons close enough together to gain the maximum energy from the attractive nucler force. At this point we begin to lose energy by fusion and would gain energy by fission. Nuclear physicists who study the structure and the energy balance in the nuclues now understand how this occurs. The change occurs at the iron nucleus, which is the most stable nucleus in nature. We can gain energy by building up iron nuclei from smaller nuclei. We can also gain energy by breaking up larger nuclei into iron. Iron contains 26 protons and its most common isotope, iron-56, has 30 neutrons.

The world is full of nuclei both much lighter and much heavier than iron. This is because of the Coulomb barrier, which prevents fission and fusion from occurring under normal conditions. To turn lighter nuclei into iron or to break heavier nuclei up into iron requires pumping in enough energy to overcome the barrier. To obtain useful energy we need a mechanism for pumping this energy in and a mechanism for using the energy that comes out. So far this has succeeded only in fission of the heaviest known nuclei like uranium, thorium and plutonium and the fusion of the lightest known nuclei like hydrogen, helium and lithium.

In uranium fission the energy released by the two pieces or fission fragments moving rapidly apart is easily turned into heat. The main problem in making a nuclear reactor is how to pump in the energy needed to get the uranium nucleus over the barrier; how to get two parts of the nucleus far enough apart so that the repulsive electric force overcomes the attractive nuclear force and the nucleus breaks apart. The nucleus has been compared to a tiny liquid drop. It normally has a spherical shape, but if it is hit in the right way it can be deformed into an egg shape. Once the deformation is large enough, the two ends of the egg are far enough apart so that they no longer feel the strong nuclear force and the repulsive elctric force makes the egg longer until it breaks apart. It was found that a neutron moving slowly past a nucleus of uranium-235 is attracted to it by the strong nuclear force to make uranium-236. But the strong forces between the neutron and the rest of the nucleus pull the nucleus into an egg shape and may deform the nucleus to the point where the two ends of the egg are pushed apart by the electrical forces and fission occurs. This was the process discovered by Hahn and Strassman in 1936.

But where can one get the neutrons to trigger off the fission of uranium? The breakthrough that made possible the exploitation of nuclear fission energy was the discovery that when uranium breaks up into two fragments like barium and krypton there are also a few neutrons emitted, two or three on the average. If one of these neutrons can hit another uranium nucleus and cause fission, we can generate what is called a "chain reaction" and the uranium fuel can be "burned up", its energy turned into heat. It is like an ordinary coal fire, where it is necessary to heat up the coal in order to start a fire burning. But once the fire is started, a "chain reaction" occurs when the heat produced by one burning coal warms another coal up to the temperature where it starts to burn, etc.

Life is not quite so simple. The neutrons emitted when a uranium nucleus breaks up are moving very fast and will pass by other uranium nuclei much too quickly to exert the strong forces that split the nucleus. The neutrons must first be slowed down by collisions with nuclei of other materials. But most other nuclei simply grab any neutron around and make a bigger nucleus. It is necessary to find nuclei which do not attract neutrons very strongly so that the neutrons can bounce around in the material, colliding with one nucleus after another and slowing down in the process, without being captured by the nucleus. The best nuclei for slowing down neutrons are those of carbon, oxygen and heavy hydrogen. The first nuclear reactors were made with uranium rods embedded in either graphite or heavy water.

Another problem is that the neutrons can escape from the reactor and be lost. The bigger the reactor, the better the chance that a neutron will be slowed down by collisions in the graphite or heavy water and eventually produce uranium fission, rather than escaping from the reactor and getting lost. If the reactor is too small, most of the neutrons will escape and there will be no nuclear chain reaction. If the reactor is big enough so that one neutron from each fission manages to make another fission, then the chain reaction goes, and the reactor is said to have reached the "critical size". The term "critical size" has now come into general use in many areas. This is the way the term was first used.

Nuclear technology has now reacahed the point where many other materials are used to slow down the neutrons, including ordinary water, and "enriched" uranium is used in which the uranium-235 isotope is concentrated to be much more than the 0.7% found in natural uranium. When pure uranium-235 is separated from natural uranium (which is mainly uranium-238) it is much easier to make a chain reaction go, and it is not necessary to slow down the neutrons. There are some "fast reactors" which use nearly pure uranium-235 and fast neutrons. A fission bomb is such a fast reactor. To make a bomb it is necessary to get the reaction going so fast that the energy is all released before the bomb blows apart.

A nuclear reactor which uses slow neutrons cannot explode like a bomb. If the reaction gets out of control and say two neutrons from each fission produce more fissions, the heat generated is so intense that it immediately damages the reactor, neutrons get lost in empty spaces and the reaction stops. It is as if someone explodes a coal fire by sending the coals out in all directions. Once the coals no longer have their neighbors to keep them hot they gradually stop burning and go out. A nuclear reactor out of control is like an overheated coal fire. It can do much damage, but it will not explode.

Getting energy out of fusion is much more difficult. Here the problem is how to pump the energy into two heavy hydrogen nuclei so that they will overcome the electric force that pushes them apart and come together to make a helium nucleus. The only way known to give them this energy is to heat up a sample of heavy hydrogen nuclei to the point where they are moving fast enough to come together against the electric force. But this requires exceedingly high temperatures. Such high temperatures are found in the interior of the Sun and in stars. Astrophysicists believe that it is this fusion energy of hydrogen turning into helium that gives the energy radiated from the Sun.

On the Earth it is much more difficult to reach the temperatures needed for fusion, and there is the problem of designing a furnace capable of withstanding such temperatures. The tokamac invented by Andrei Sakharov is one idea for how to contain the hot fusion fire without destroying the furnace. There are also other ideas. But we still have a long way to go before the problems of how to pump the energy into hydrogen nuclei are solved in a practical way and fusion energy becomes available on a commercial level.

The Nuclear Glue

As the nuclear physicists learned more and more about the way the neutrons and protons fit together to make nuclei, they also attempted to understand the nature of the forces, the nuclear glue, that holds the nucleus together. They originally believed that the same ideas which worked so well for unraveling the structure of the atom would also give a successful description of the structure of the nucleus. They developed a model, called the nuclear shell model, in which the neutrons and protons moved in orbits inside the nucleus very much like the orbits of the electrons in atoms outside the nucleus. The motion of the neutrons and protons was determined by the two kinds of forces acting on them, the repulsive electric force between the protons and the new attractive nuclear force which acted only at very short distances.

The remaining open question was the origin of this new short range "nuclear glue". The physicists first believed that it must come from another field of force, different from electricity, but described in the same way that Maxwell described electromagnetism. There would be new particles emitted from nuclei, analogous to the Einstein's photons of light. The Japanese physicist Yukawa proposed this idea in 1935 and called the new particle the meson. It would be about 200 times heavier than the electron and yet much lighter than the proton which is 2000 times heavier than the electron. In 1937 a new particle 206 times heavier than the electron was found in experiments detecting cosmic rays coming from outer space, and physicists became very excited. They were sure they had found Yukawa's meson and were close to solving the riddle of the nuclear glue. When I was a graduate student in the late 1940's many of my fellow students were examining different possible versions of Yukawa's meson theory of nuclear forces as their Ph.D. thesis research projects in nuclear physics. Some theories were exactly analogous to electromagnetism and had two kinds of forces, electric and magnetic; others had only one kind of force. There were also different possibilities for the internal spin of the meson. The photon has a spin twice that of the electron; there could be mesons with this spin and also mesons with no spin. Meanwhile the experimenters were trying to learn as much as they could about the meson from the only source that was available, the cosmic rays from outer space. High energy accelerators were being built which would later produce these mesons. But at that time, when Western technology was still recovering from World War II, accelerator development was still in its infancy.

Then the trouble started. The mesons in the cosmic rays did not behave like nuclear glue at all. They traveled right through large amounts of matter and did not seem to be affected by the strong forces that should have been acting on them. It soon became clear that these particles were not Yukawa's mesons at all. They behaved exactly like electrons, which did not have any strong nuclear forces. We now call this particle the muon or mu (μ) meson. It had no place in the nuclear physicists scheme of things, and was aptly described by Rabi's famous remark "Who ordered that?"

Physicists became excited again with the discovery of a new meson about 280 times heavier than the electron which did interact strongly with neutrons and protons and looked like it was the nuclear glue. This meson was called the pion or pi meson (π). Soon there were theories about how the pion was emitted and absorbed by nuclei, just like light photons are emitted and absorbed by atoms. But this euphoria did not last long either. The experimentalists kept on discovering more and more particles, filling out the tables shown in the previous chapter. There were too many mesons. The pion had no spin. There were other mesons with the same spin as the photon. There were the new strange particles, and new strange mesons as well. All these mesons could not be the nuclear glue, and there was no reason to pick one over the other.

Now we believe that all the mesons and baryons are made of smaller objects called quarks, and that the basic strong force that holds nuclei together must be found at a deeper level. A theory called "Quantum Chromodynamics" (QCD), a beautiful generalization of Maxwell's electrodynamics, describes the very strong forces between quarks that glue them together to make mesons, neutrons, protons and the other baryons. To understand the deeper level where we have a structure that looks like atoms and molecules on a smaller scale, we have to go beyond the models of neutrons and protons making nuclei and begin with quarks making neutrons and protons. It is the neutrons and protons themselves that are analogous to atoms. The nuclei are more like molecules.

The electrons and nucleus of the atom are held together with the forces of electromagnetism. These same electric forces also act between atoms and bind them together to make molecules like water or carbon dioxide. When two hydrogen atoms, each with one proton and one electron are far apart, there is no electric force between them. Each positively charged proton is repelled by the positively charged proton in the other atom and attracted by the negatively charged electron in the other atom. The same is true for the electrons in each atom which are repelled by the electron in the other atom and attracted by the proton. These two forces depend upon the distance between the two particles involved, and cancel exactly when the distances are the same. The electron in one atom moving around the proton is sometimes closer to the other atom and sometimes farther away. But on the average it is the same distance from the other atom as the proton and the forces cancel.

When the two atoms are very close together this picture breaks down because each electron can now move in orbits around both protons instead of each electron around its own proton. This is the way molecules are formed. Before this interplay of forces was understood, it was simply observed that there was no force between atoms at large distance, but that peculiar forces existed at very short distances when the distance between atoms was comparable to the size of the individual atoms.

We now believe that this two-step description repeats itself with the new strong force of quantum chromodynamics (QCD). The particles of this new field analogous to the photons of electromagnetism are called "gluons" because they provide the real nuclear glue. The forces of the gluon field bind quarks into neutrons and protons like the forces of the electromagnetic field bind electrons and nuclei into atoms. When two neutrons or protons are far apart, this force cancels out in the same way as the force between two atoms. It is only when the distance between neutrons and protons is comparable to their size that there is also a force between them. This force which acts only at very short distances binds them into nuclei just like the force between atoms binds them into molecules. So nuclei have turned out to be the molecules of the deeper level of force, not the atoms.

One remaining puzzle was the weak force which was responsible for the phenomenon of radioactivity. A neutron left alone does not live forever; it decays into a proton, and electron and a neutrino. Neutrons and protons in a nucleus also can turn into one another by emitting electrons or positrons and neutrinos. It was very tempting to find a new theory which would describe these transitions like the transitions in which photons and gluons were emitted by atoms and nuclei. But many years passed between the discovery of radioactivity and its explanation by a theory with a new field of force.

7

The Weak Force and the Discovery of the W Particle

The discovery of two new particles called W and Z was reported by CERN, an international High Energy Physics laboratory in Geneva, Switzerland run by all the Western European countries. These experiments are so enormous and expensive, that no single university or country smaller than a superpower can afford them. There are other laboratories like this only in the United States and Russia (the former Soviet Union).

The W has been called the "carrier of the weak force". What does this mean? How does a particle carry a force? What is this weak force?

The discovery of the W has provided the missing piece in the puzzle of radioactivity, which has been known and used since the beginning of the twentieth century without any real understanding of what it is. The pioneering investigations by Pierre and Marie Sklodowska Curie into why photographic film became exposed when placed near uranium led to the discovery of many materials that emitted a new kind of invisible radiation called beta rays. They passed through material like X-rays and affected film, but were also very different from X-rays. The process of emission of beta rays is called radioactivity. Further investigations produced a catalogue of over a thousand materials called radioactive or radioisotopes which emit beta rays. But the puzzle of what new unknown force makes the beta rays come out of these materials remained unsolved. The force must be very weak because it produces results much more slowly than other known forces like electricity and magnetism.

This weak force is both a blessing and a curse. It gives us the radioisotopes that have proved to be so useful in medicine and industry. But it also gives dangerous radioactive fallout from nuclear bomb tests and radioactive waste products from nuclear reactors. The weak force has given mankind the radiation therapy that can cure cancer together with the radioactive fallout that can cause cancer. Radioactivity can be slow, often very slow, and its speed cannot be controlled from outside like an X-ray machine. Each material emits its own particular kind of beta rays, with its own energy and speed or lifetime. The energy stored in some radioactive materials is released as beta rays in a few minutes or a few hours or days. Others may release their energy very slowly over periods of millions or billions of years. We know how to make these materials in the laboratory. But once they are made, they release their energy in their own way at their own speed. We do not know any practical way to control or change it.

Radioactivity can be compared with the light radiated from fluorescent materials used to paint light switches, faces of watches and clocks, and other objects where a tiny amount of light can be usful in dark places. These materials glow for a short time and lose their optical energy after a few minutes or hours. But some radioactive materials continue glowing for millions of years. Man has already found many uses for radioactive glowing materials. Some radioisotopes which glow for a few days are used in medical research, diagnosis or therapy. Research scientists use them to study how elements like iron, iodine, calcium or potassium are used in a healthy animal, plant or human being, and how this is changed by diseases. They inject or feed tiny amounts of radioisotopes of these elements and follow them through the body by detecting the radioactive glow from outside or from blood or other fluid samples. A few radioisotopes immediately go to certain types of tumors, where their glow can reveal the location of the tumor on diagnostic x-ray film or may even kill cancer cells directly. These short-lived radioisotopes disappear in a few days after their work is finished and cause no permanent damage.

Longer lived radioisotopes, like carbon-14 which glows for about 6000 years, are useful for dating in geology and archeology. Cosmic rays coming continuously from outer space convert tiny amounts of ordinary nitrogen atoms in the atmosphere into radioactive carbon-14, which glows very weakly but for a long time. This tiny amount of radiation does no harm, but sensitive instruments can detect it. When scientists find this glow in rocks, bones of animals or sea shells buried in the earth where cosmic rays cannot penetrate, the glow tells them that these objects were once on the earth's surface, perhaps thousands of years ago, where they were in contact with the atmosphere and picked up the radioactive carbon atoms from the air. If there is no glow at all, they must have been buried for many thousands of years. Precise measurements of how much glow is still left can tell the scientist how long ago these objects lost contact with the atmosphere.

The radioisotopes left glowing after fuel has been burned up in a nuclear reactor are like the glowing coals left after a fire, except that nobody knows how to extinguish them nor how to speed up the process and burn them up quickly. The glow dies away at its own speed with no respect for our desires. This radioactive waste created in the operation of any nuclear power plant must somehow be safely stored somewhere.

What is this radioactive glow? Why does it last for such a long time, burning so much slower than ordinary electrical, chemical and optical glows? These questions have puzzled scientists for a long time. We know now that electrical, magnetic, chemical and optical glows are all described by a unified theory called electromagnetism originally developed in 1864 by the British physicist Maxwell to explain electricity and magnetism but which has turned out to explain very much more.

Is radioactive glow another aspect of electromagnetism or is it completely different? Physicists who followed Einstein's approach tried to describe radioactivity like the other glows and had to find a reason why it was so much slower. The slowness might be caused by insufficient energy, like a fire which does not start because we haven't put in enough heat. Even in ordinary fire, some materials burn much faster than others. Much more energy is needed to start a wood fire than a paper fire. Very much more energy and a very high temperature is needed to make iron glow and to burn rapidly into rust. At ordinary temperatures, iron rusts very slowly. The radioisotopes could be like iron, "burning" or glowing very slowly at ordinary temperature, and requiring much higher temperature to be speeded up. The temperature needed is impossibly high, millions of billions of degrees Celsius. Such temperatures may have been present at the origin of the universe, but nobody can make them now on earth.

In this unified approach following Einstein, radioactivity results from an energy transfer between transmitting and receiving antennas both deep in the same atom. The energy is carried between them by a new particle called the W which never comes out of the atom. At the receiving antenna the W disappears and its energy is transformed into beta rays, just like a beam of sunlight is stopped by our bodies and the energy is transformed into heat. The process goes slowly at normal temperatures, much slower than iron rusting, because tremendous energy is needed to make a W. This slow process is uncontrollable. There is no way to coat or otherwise protect the radioisotopes against radioactivity in the way that iron can be protected from rusting, because radioactive processes do not depend upon anything like oxygen from outside. Everything needed is already inside their atoms, and the glow cannot be stopped without getting inside the atom.

How can we test this theory when we cannot heat even a tiny amount of material like a millionth of a gram to the temperature needed to make the W's and Z's and get them out of the atom? A millionth of a gram still contains about a billion atoms. Maybe we can pump enough energy into one or two atoms to heat them to the point where one W comes out. Physicists have been building machines which shoot two atomic nuclei at one another at very high speeds and have been looking to see whether the collision produces enough energy to make the W. Over the past twenty years, bigger and bigger machines have been built, but no W's were found. Each time, some physicists concluded that more energy was needed to make the W, while others insisted that the whole unified theory of the weak force was wrong and that there was no W.

Meanwhile new experiments gave more information about the weak forces of radioactivity. A new and more refined unified field theory by the physicists Glashow, Weinberg and Salam put the weak force and electromagnetism together in a very beautiful way, with the W particle playing the same role for the weak force as Einstein's photon for electromagnetism. They not only explained everything that was known, but also told exactly how much energy was needed to make the W. It was more energy than any previous machine could give, but was within reach of modern technology. Glashow, Weinberg and Salam received the Nobel Prize for physics and the CERN laboratory in Geneva built a machine to look for the W.

The new machine made the W with exactly the right energy predicted by Glashow, Weinberg and Salam. Their unified theory of radioactivity, electricity and magnetism has been confirmed. This step in the direction of Einstein's goal brings one chapter in our understanding of the forces of nature nearly to an end and perhaps opens a new chapter.

What are all these particles? Who needs them? These questions lead us back to the history of the long search for the elements from which all matter is made. In the ancient times the elements were believed to be earth, air, fire and water. Today we know that these are not elements. The chemists found water to be a compound that can be broken up into hydrogen and oxygen. They called hydrogen and oxygen elements because they could not break them up by chemical techniques. They believed that these chemical elements were made of indivisible atoms.

But new experiments broke up these atoms into electrons and nuclei. Scientists had found one simple method to test whether particles were indivisible or made of smaller building blocks. Just hit anything hard enough and see if it breaks up. Further experiments broke up nuclei into neutrons and protons. But new experiments banging two protons together at very high speeds failed to break up the proton. The enormous energy of the high speed protons was converted into new matter, and many new particles were created in the collision, but the protons remained intact.

Inside the nucleus there were two kinds of particles: the proton, which carries electric charge, and its partner the neutron, which has no charge. Outside the nucleus there was the electron, which carries an electric charge, and it was found also to have a partner called the neutrino, which has no charge. Protons, neutrons, electrons and neutrinos have remained unbroken. In the 1940's scientists believed that they had found the real elements from which all matter is made. They were this "family" of particles with two couples, one inside the nucleus and one outside.

But soon new unbroken particles were discovered. The first of these, now called the muon, was very much like the electron, only much heavier. Nobody understood why there was such a particle. It is not found in normal matter, but is created by the energy released in high speed collisions between protons. It lives only for a millionth of a second and then decays into an ordinary electron and two neutrinos. Soon more particles were found which were made in high energy collisions which lived only for a few billionths of a second. There were soon too many new particles to fit into any sensible theory. This dilemma has consistently troubled scientists who search for a way to build all matter out of a few elements. At each stage when scientists think that they have found the elements, more and more are discovered and there soon are too many.

Scientists do not like to believe that there are a large number of elements, all different and completely unrelated. They look for an explanation at some deeper level. Today we know that the neutron and proton are not elements, but are made of two kinds of smaller objects called quarks. Again there is a couple, which have been given the curious names of "up" and "down" quarks. The electron and neutrino are still elementary, so there is again a family of four with two couples, the up and down quarks, and the electron and its neutrino. This family would explain everything.

But there was still this muon, a heavy electron which was extra. Like the electron and neutrino, these particles were always found outside the nucleus. There were also new strange particles found inside the nucleus, which could not be made from up and down quarks. A third quark was needed. It has been called the "strange quark." For a while everyone thought that this was the end. There were all kinds of theoretical reasons why three quarks seemed to be the right number, and there were now three kinds of quarks inside the nucleus and three kinds of particles outside, electrons, muons and neutrinos.

But then the muon was found also to have a partner, a muon-neutrino which was different from the neutrino partner of the electron. The fascinating experiment which found the muon neutrino was carried out by a group of physicists at Columbia University, which included Leon Lederman. Now there seemed to be no relation between the triplet of particles inside the nucleus and the quartet outside.

Here again the searches for the building blocks of matter and the forces that hold the blocks together are intertwined. An explanation of one seems to create puzzles for the other.

We now return ro the new theory which explained radioactivity and predicted the existence of the W particle. This theory needed another new, not yet discovered building block. It needed two families of four basic particles, each with two quarks found in the nucleus and four particles found outside. They predicted that the "strange" quark must also have a partner, which they called the "charmed" quark, and that the strange and charmed quarks formed a second family, or a "second generation" of particles together with the muon and its neutrino partner.

In the summer of 1974 particle physicists were very confused. The charmed quark had not yet been discovered, and most physicists did not believe that it existed at all. In the following year the charmed quark was discovered with new puzzles leading to the discovery of more new building blocks.

Once again there were still too many particles and the theorists are trying to explain why there are so many. There are still two competing approaches focusing on building blocks or forces. One considers particles as compounds, made of a much smaller number of even tinier building blocks not yet discovered. The large number of chemical elements and later the large number of different nuclei were explained by showing them to be compounds. The other approach looks for a unified description of all the different forces of nature, like the unified field theory that Einstein tried to find in his later years. Electricity, magnetism, radio waves, microwaves, light waves, infra-red radiation, ultraviolet radiation, nuclear gamma radiation and X-rays have been unified in this way. They are now all described as radiation of the same kind of particles called photons. The recently discovered W particle brings the weak force of radioactivity into this unified family of forces. Perhaps all the generations of particles are all different aspects of the same particle, like all the photons. Nobody knows the answers to these questions. That's what makes the puzzle so interesting.

The strong force which holds atomic nuclei together and the force of gravity were not included in the unified theory that combined radioactivity, electricity and magnetism. A new theory called "quantum chromodynamics" (QCD) was developed to include the strong force which binds quarks into particles. Thie new theory has been called the "Standard Model" for particle physics. All experimental results in particle physics were then explained by this standard model.

Theorists continue Einstein's search for unification by developing new theories which go beyond this known standard model and also include gravity. Again searches for new forces have led to searches for new undiscovered building blocks. One theory, called "super-symmetry", requires all known particles to have new partners in new supersymmetric families. So far none of these new "superparticles" have been found but there are active searches at all accelerators.

New machines at even higher energies are being built to test these theories, and perhaps to make more exciting new discoveries not predicted by the theorists.

The most ambitious new project is the "Large Hadron Collider" (LHC) built in Geneva and very much bigger than the machine which discovered the W. Hopefully they will make even more exciting discoveries.

The latest exciting discovery in 2012 is the finding at the new LHC of a particle called the Higgs boson. The Higgs theory claims that their boson is associated with a new field of force in the same way that the weak force is associated with the W particle. This Higgs field force is supposed to gives all particles their masses. But whether this is really true still remains to be seen. The Higgs theory does not explain why there are so many different particles and building blocks and why their masses are different. This puzzle still remains a challenge for theorists.

What is matter made of? The more questions we answer in our search for the elementary constituents of matter, the more new questions we find. The more we learn, the more there seems to be left to learn.

Index

accelerators, 86, 90, 92
Alexeyeva, Lisa, 10, 28
American Physical Society, 6
Andrei Sakharov Gardens, 27
Andrei Sakharov Prize, 26
antimatter, vii, 22, 72, 73, 76, 110

baryon decuplet, 107, 110
baryons, 33, 34, 37, 43, 101, 105, 106, 134
Bhabha scattering, 78–80
big bang, 22
Bonner, Elena, 13, 28

CERN, 137, 141
charmed quarks, 39, 42, 43, 113, 114, 144
Chicago Sun-Times, 6, 9
color change, 39, 88
Creationism, 82
cyclotron, 79

Dead Sea, 126
deuteron, 128
Dirac, P.A.M., 70

Eightfold Way, 104
electric charge, 30, 35, 37, 72, 87, 95,
    98–102, 104, 105, 109–111, 123, 124,
    128, 142
electromagnetic induction, 122
electromagnetism, 121
electron neutrino, 112
electronics, 52, 53, 56
electrons, 36, 37, 40, 71, 76–79, 85, 87, 98,
    142
evolution, 80–82

Fermi National Accelerator Laboratory
    (Fermilab), 115
Fermi statistics, 38, 39
fission bomb, 131

gamma rays, 73, 74, 78
Gell-Mann, Murray, vii, 37, 87, 101, 104
genetic code, 51
gluons, 135
gravity, 121

Higgs boson, 145
hydrogen bomb, vii, 6, 13, 19, 20

Il Tempo, 3
independent convergence, 49, 50
isotopes, 98, 99

Kriegspiel, 55, 56

lambda, 40–42, 44, 45
language code, 51, 52, 57–60
Large Hadron Collider, 145
leptons, 46, 112, 119
liquid drop, 130
LITAF program, 61, 62

Mössbauer effect, 69
matter, vii, 22, 76
matter–antimatter, 23
Maxwell's equations, 53
Mediterranean Sea, 126
Mendeleev's table, 36, 37, 96, 97, 104, 111,
    119

meson "nonet", 111
mesons, 33, 34, 37, 41–43, 50, 101, 110, 133, 134
microwave radar, 53, 54
Moscow Seminars in Exile, 8, 9
Mott Scattering, 77, 78
muon, 111, 113, 134, 142, 143
muon neutrino, 112

Neptune, 121
neutrino, 111, 142
neutron, x, 22, 32, 36, 37, 43, 44, 50, 98, 100, 105, 124, 134, 142
neutron magnets, 100, 101, 109
New Scientist, 7, 8
New York Times, 6, 7, 8, 112
Nobel Peace Prize, 19
November revolution, 112, 113, 144
nuclear fission, 125, 127
nuclear shell model, 133

octet
     baryon octet, 102, 104, 108, 110
     meson octet, 102, 104, 110, 111

pion, 134
positron emission tomography (PET), 73, 74
positrons, 72–74, 76–79, 87
preons, 34
proton decay, 13, 14, 16, 23
proton magnets, 100, 109
protons, x, 22, 32, 36, 37, 40, 43, 44, 50, 88, 95, 98, 100, 105, 124, 134, 142
pseudoscience, 47, 48, 67

Quantum Chromodynamics (QCD), 31, 39, 40, 42, 45, 134, 135, 144
Quantum Electrodynamics (QED), 39
quark magnets, 108
quarks, x, 16, 21, 34–38, 43–46, 48, 85–88, 104–107, 112, 116, 119, 134, 143
     bottom quarks, 39, 114, 115, 119
     charmed quarks, 39, 42, 43, 113, 114, 144

down quarks, 38, 40, 41, 43, 106, 108, 109, 111, 143
strange quarks, 38, 40–43, 101, 106, 108, 111, 113, 143
top quarks, 39, 114, 115, 119
up quarks, 38, 40, 41, 43, 106, 108, 109, 111, 143

refuseniks, 4
rishons, 34
Rutherford scattering, 77, 94, 95

Sakharov, Andrei, vii, viii, 1–17, 19–28, 30–32, 34, 35, 40, 42, 44–46
Sakharov, Orlov and Sharansky (SOS), 5, 11
scientific establishment, 68, 69
Semyonov, Alexei, 12, 13, 23
Sharansky, Anatoly, 10, 11
social science, 47, 48, 50
special relativity, 65, 70–72, 75, 81
spin, 101, 107, 110
Stanford Linear Accelerator Laboratory, 118
strong nuclear force, 124, 125, 128

tau, 113, 114, 119
tau neutrino, 119
telescope, 66, 67
three-generation model, 115, 119
tokamak, 21

upsilon, 114, 115
Uranus, 121

Voice of America, 5, 7

W particle, 116, 137, 141, 143–145
Washington Post, 4, 6, 14
weak force, 136–138, 141
Weizmann Institute, 2, 4, 14, 19, 21, 25–27, 29, 30, 35, 44, 46, 70, 73, 118

Z particle, 137
Zweig, George, vii, 104